ANNALS *of* THE NEW YORK ACADEMY OF SCIENCES

The New York Academy of Sciences
7 World Trade Center
250 Greenwich Street, 40th Floor
New York, NY 10007-2157

annals@nyas.org
www.nyas.org/annals

Published by Blackwell Publishing
On behalf of the New York Academy of Sciences

Boston, Massachusetts
2012

ANNALS *of* THE NEW YORK ACADEMY OF SCIENCES

VOLUME 1270

ISSUE

Thymosins in Health and Disease II

The Third International Symposium

ISSUE EDITORS

Allan L. Goldstein[a] and Enrico Garaci[b]

[a]The George Washington University School of Medicine and Health Sciences
[b]Istituto Superiore di Sanità

TABLE OF CONTENTS

vii Introduction for *Thymosins in Health and Disease*
Allan L. Goldstein and Enrico Garaci

Thymosin α1: immunomodulation, immunopharmacology, infectious diseases, and cancers

1 Thymosin α1: a novel therapeutic option for patients with refractory chronic purulent rhinosinusitis
Virgil A.S.H. Dalm, Harm de Wit, and Hemmo A. Drexhage

8 Thymosin α1 in melanoma: from the clinical trial setting to the daily practice and beyond
Riccardo Danielli, Ester Fonsatti, Luana Calabrò, Anna Maria Di Giacomo, and Michele Maio

13 Thymosin α1 as a stimulatory agent of innate cell-mediated immune response
Annalucia Serafino, Pasquale Pierimarchi, Francesca Pica, Federica Andreola, Roberta Gaziano, Noemi Moroni, Manuela Zonfrillo, Paola Sinibaldi-Vallebona, and Enrico Garaci

21 Thymosin α1 continues to show promise as an enhancer for vaccine response
Cynthia Tuthill, Israel Rios, Alfonso De Rosa, and Roberto Camerini

Clinical and preclinical applications of thymosin β4: wound healing, eye injuries, cardiovascular diseases, and neurological injuries

28 The use of angiogenic-antimicrobial agents in experimental wounds in animals: problems and solutions
Paritosh Suman, Harikrishnan Ramachandran, Sossy Sahakian, Kamraan Z. Gill, Basil A. J. Horst, Shanta M. Modak, and Mark A. Hardy

37 The regenerative peptide thymosin β4 accelerates the rate of dermal healing in preclinical animal models and in patients
Terry Treadwell, Hynda K. Kleinman, David Crockford, Mark A. Hardy, Giorgio T. Guarnera, and Allan L. Goldstein

45 Thymosin β4: a potential novel dry eye therapy
Gabriel Sosne, Ping Qiu, George W. Ousler 3rd, Steven P. Dunn, and David Crockford

51 Neuroprotective and neurorestorative effects of thymosin β4 treatment following experimental traumatic brain injury
Ye Xiong, Asim Mahmood, Yuling Meng, Yanlu Zhang, Zheng Gang Zhang, Daniel C. Morris, and Michael Chopp

59 Use of the cardioprotectants thymosin β4 and dexrazoxane during congenital heart surgery: proposal for a randomized, double-blind, clinical trial
Daniel Stromberg, Tia Raymond, David Samuel, David Crockford, William Stigall, Steven Leonard, Eric Mendeloff, and Andrew Gormley

66 Cardiac repair with thymosin β4 and cardiac reprogramming factors
Deepak Srivastava, Masaki Ieda, Jidong Fu, and Li Qian

Late breaking paper session

73 NMR structural studies of thymosin α1 and β-thymosins
David E. Volk, Cynthia W. Tuthill, Miguel-Angel Elizondo-Riojas, and David G. Gorenstein

79 Fragments of β-thymosin from the sea urchin *Paracentrotus lividus* as potential antimicrobial peptides against staphylococcal biofilms
Domenico Schillaci, Maria Vitale, Maria Grazia Cusimano, and Vincenzo Arizza

86 Development of an analytical HPLC methodology to study the effects of thymosin β4 on actin in sputum of cystic fibrosis patients
Mahnaz Badamchian, Ali A. Damavandy, and Allan L. Goldstein

93 The role of biologically active peptides in tissue repair using umbilical cord mesenchymal stem cells
Carlos Cabrera, Gabriela Carriquiry, Chiara Pierinelli, Nancy Reinoso, Javier Arias-Stella, and Javier Paino

98 Identification of interaction partners of β-thymosins: application of thymosin β4 labeled by transglutaminase
Christine App, Jana Knop, Hans Georg Mannherz, and Ewald Hannappel

105 Antibodies in research of thymosin β4: investigation of cross-reactivity and influence of fixatives
Jana Knop, Christine App, and Ewald Hannappel

112 Thymosin β4 sustained release from poly(lactide-co-glycolide) microspheres: synthesis and implications for treatment of myocardial ischemia
Jeffrey E. Thatcher, Tré Welch, Robert C. Eberhart, Zoltan A. Schelly, and J. Michael DiMaio

---

120 Corrigendum for Ann. N.Y. Acad. Sci. 2005. 1051: 779–786

121 Corrigendum for Ann. N.Y. Acad. Sci. 2012. 1254: 57–65

Ann. N.Y. Acad. Sci. ISSN 0077-8923

ANNALS OF THE NEW YORK ACADEMY OF SCIENCES
Issue: *Thymosins in Health and Disease*

Introduction for *Thymosins in Health and Disease*

The Third International Symposium on Thymosins in Health and Disease brought together many of the leading scientists, clinicians, and thought-leaders from North and South America, Europe, and Asia to discuss the latest advances and clinical applications of the thymosins in both basic and clinical areas. The symposium, held in Washington, DC on March 14–16, 2012, was sponsored by the George Washington University, the Istituto Superiore di Sanitá (ISS), and the University of Rome "Tor Vergata."

Of the six plenary sessions, three were devoted to advances in basic research and two were devoted to translational advances in the broad areas of immunization, immune deficiency diseases, infectious diseases, cancer, wound healing, and treatment of eye injuries, cardiovascular diseases, and neurological injuries. A sixth, late-breaking, basic research session focused on advances in nuclear magnetic resonance analysis of thymosin peptides, novel antimicrobial agents, use of stem cells and growth factors for tissue and burn repair, and advances in analysis of peptides.

Luigina Romani, professor of microbiology at the University of Perugia, and Deepak Srivastava, director of the Gladstone Institute of Cardiovascular Disease at the University of California, San Francisco, gave keynote lectures and were co-recipients of the 2012 Abraham White Distinguished Science Award, which was presented at the Awards Banquet. The 2012 Abraham White Distinguished Humanitarian Award was given posthumously to Michael Stern and was accepted by his daughter Margaret Stern. Mr. Stern was recognized for his lifetime of good deeds aimed at supporting and improving Italian–American scientific relations and for his journalistic excellence and philanthropic endeavors. Stern played an important role in co-creating the Fisher Center for Alzheimer's Research Foundation at the Rockefeller University and in starting the Michael Stern Parkinson's Research Foundation.

Dr. Romani was honored for her pioneering studies and scientific contributions, which have significantly increased our understanding of the molecular mechanism of thymosin α1 (Tα1), and for elucidating the important role it plays in modulating Toll-like receptors and in immunity. Her breakthrough discovery of tryptophan metabolites as potential antifungal strategies, and her most recent report of functional genomics to predict risk for fungal infections in transplant patients have provided the biomedical community with new strategies for the treatment of serious fungal diseases.

Dr. Srivastava was honored for his pioneering studies and scientific contributions, which have significantly advanced our understanding of the role of thymosin β4 (Tβ4) in the development and functioning of the human heart. His studies have provided a novel therapeutic target in the setting of acute myocardial damage. His most recent studies on the molecular events regulating early and late developmental decisions that instruct progenitor cells to adopt a cardiac cell fate and, subsequently, to fashion a functioning heart hold much potential in preventing congenital defects and treating acquired heart disease, particularly with cardiac-specific differentiation of embryonic stem cells. Most recently, he has been able to reprogram adult fibroblasts directly into cardiomyocyte-like cells for regenerative purposes.

doi: 10.1111/j.1749-6632.2012.06778.x

The conference scientific summary

The papers included in the two volumes of these proceedings highlight the progress that has occurred in the biological, chemical, and clinical applications of the thymosins in health and disease. Volume 1269 (plenary sessions I, II, and III) includes papers that summarize the most recent advances in basic research, which have provided the scientific foundation for translational studies with several of the thymosin peptides. Volume 1270 (plenary sessions IV, V, and VI) includes papers on clinical applications of thymosins α1 and thymosin β4. This volume also has a number of reports describing novel labeling techniques, formulations, and diagnostics, as well as the characterization of a thymosin β4 fragment with antimicrobial properties isolated from the sea urchin *Paracentrotus lividus*.

Plenary session I focused on advances in our understanding of the family history of the β-thymosins and the chemistry and role of prothymosin α (ProTα), Tα1, and Tβ4 on cellular receptors, signaling, and multifunctionality. John Edwards gave a brilliant lecture on the origin of the β-thymosin family, tracing its evolution from single-celled organisms to fish and land vertebrates. Milica Radisic reported on a significant advance in the engineering and encapsulation of Tβ4 in a collagen–chitosan hydrogel, which allows for controlled release of Tβ4 at nearly zero order kinetics. The presentations by Francesca Fallarino and Enrico Garaci focused on the interface between Tα1 and Toll-like receptors, and the role of Tα1 in enhancing antitumor activity. In his presentation, Yeu Su focused on upregulation of Tβ4 in a colon cancer model and the use of Tβ4 as a potential therapeutic marker for human colon cancers. In the last plenary talk of the session, Hiroshi Ueda presented new evidence that ProTα is a multifunctional nuclear protein that has important activities in the central nervous system.

Plenary session II focused on advances in our understanding of key molecular markers and the activity of Tβ4 in accelerating wound healing, reducing fibrosis, and inflammation in injured tissues. Gavino Faa began the session with a talk covering the general expression of Tβ4 in human tissues. Karina Reyes-Gordillo discussed Tβ4 and the mechanism by which it prevents liver fibrosis in a rat model using carbon tetrachloride. Enrico Conte showed us through his *in vivo* studies that, in the lung, Tβ4 protects from bleomycin-induced damage in C57BL/6 mice. Sprague Hazard's presentation provided new evidence that the myofibroblast plays an important role in scar formation and the mechanism by which Tβ4 reduces scar formation following injury. In the final presentation of the session, Hee-Jae Cha discussed the stabilization of the hypoxia-inducible factor-1α by Tβ4 through an oxygen-independent manner.

Plenary session III focused on the activities of Tβ4 in cardiovascular protection, neuroplasticity, regeneration, and stem cell differentiation. Sudhiranjan Gupta began the session with a very informative talk on how Tβ4 protects the heart and reduces inflammation and fibrosis. Nicola Smart then discussed the facilitation of myocardial regeneration by Tβ4 via activated epicardial progenitor cells. Christoffer Stark presented data on the cardioprotective effects of Tβ4 after myocardial infarction. Christian Kupatt-Jeremais presented new data documenting the essential elements of Tβ4-mediated collateral growth. In studies of the brain following ischemic injury, Dan Morris outlined the mechanism by which Tβ4 mediated oligodendrogenesis after stroke. David Bader closed the session with a fascinating presentation showing how Tβ4 activates mesothelial cells from the omentum to repair damaged blood vessels.

Plenary session IV focused on Tα1 in areas of immunomodulation, immunopharmacology, infectious diseases, and cancer. Virgil Dalm gave the first talk of the session on the gene expression abnormalities of motility disturbed monocytes in recurrent and chronic purulent rhinosinusitis and its correction by thymosin fraction 5 and Tα1. Michele Maio then discussed the role of Tα1 in treating patients with advanced melanoma. Annalucia Serafino's presentation outlined studies documenting the role of Tα1 in innate cellular immunity. Luigina Romani, one of the keynote speakers, reported on the first clinical trial of bone marrow–transplanted patients treated with Tα1 to prevent infections. Cynthia Tuthill provided an excellent summary of the pleotropic activities of Tα1 and the activity of Tα1 as an enhancer of vaccine responses. Roberto Camerini ended the session with a presentation of the results of a late stage clinical trial of Tα1 as an adjuvant for enhancing the efficacy of the H1N1 vaccine in patients with end-stage renal disease.

Plenary session V dealt with the clinical and preclinical applications of Tβ4 and its role in wound healing, repair of eye injuries, and Tβ4's effects in treating cardiovascular disease and neurological injuries. Mark Hardy started the session by presenting the results of a study of combination therapy with Tβ4 in an animal model. He provided the first experimental evidence indicating that using Tβ4 in combination with silver sulfadiazine enhances the healing of acute and infected wounds. Terry Treadwell discussed the results of two phase II trials providing the first clinical evidence that Tβ4 accelerates the rate of healing of both pressure and venous stasis ulcers. Gabe Sosne focused his remarks on Tβ4 as a potential novel dry eye therapy. Ye Xiong discussed Tβ4's improvement of functional recovery after traumatic brain injury, as well as its neuroprotective and neurorestorative effects. Daniel Stromberg concluded the session with his presentation on the proposed use of Tβ4 as a cardioprotectant during congenital heart surgery in a randomized, double-blind clinical trial.

Plenary session VI was a late-breaking paper session including such topics as the NMR structure of human Tα1, presented by David Volk; the isolation of a novel β-thymosin from the sea urchin *Paracentrotus lividus* and its use as a novel antimicrobial peptide against *Staphylococcal* biofilms, presented by Domenico Schillaci; and Tβ4's key role in the regulation of actin polymerization and depolymerization, and the development of an analytical HPLC methodology to study actin in the sputum of cystic fibrosis patients, presented by Ali Damavandy. In additional presentations, Javier Paino presented preliminary data on the re-epithelialization of burn injuries using cultured umbilical cord cells; Christine App and Jana Knop presented data on Tβ4 labeling and molecular investigations using Tβ4 antibodies. Ending the session was a talk by Jeffrey Thatcher about implantable poly(lactide-co-glycolide) microspheres for sustained release of Tβ4.

The increasing number of thymosin peptides available synthetically has significantly accelerated animal model experimentation in the field and has helped researchers to consider novel clinical applications. These *Annals* volumes should be of keen interest to basic and clinical scientists, pharmacologists, immunologists, biochemists, and cell biologists with interests in the area of biological response modifiers. It should also be of special interest to physicians studying the clinical applications of the thymosins in both health and disease.

We would like to thank the program committee: Roberto Camerini, David Crockford, Paul Riley, Ewald Hannappel, Hynda Kleinman, Luigina Romani, Cynthia Tuthill, Deepak Srivastava, and Bruce Zetter for their help in planning this important scientific symposium. We would also like to acknowledge the editorial staff of *Annals of the New York Academy of Sciences* for their support in editing and publishing this issue.

We are very appreciative of the generous support of the following organizations: the George Washington University; the University of Rome "Tor Vergata;" University of Catania; the Istituto Superiore di Sanitá; Regina Elena Cancer Hospital; Sigma-Tau Pharmaceuticals, Inc.; SciClone Pharmaceuticals, Inc.; RegeneRx Biopharmaceuticals, Inc.; Rothwell, Figg, Ernst & Manbeck, P.C.; Fisher Clinical Services, Inc.; and PolyPeptide Laboratories, Inc. Without the support of our friends from the private sector, this type of educational meeting would not be possible.

ALLAN L. GOLDSTEIN
The George Washington University, Washington, DC

ENRICO GARACI
Istituto Superiore di Sanitá, Rome, Italy

Ann. N.Y. Acad. Sci. ISSN 0077-8923

ANNALS OF THE NEW YORK ACADEMY OF SCIENCES
Issue: *Thymosins in Health and Disease*

Thymosin α1: a novel therapeutic option for patients with refractory chronic purulent rhinosinusitis

Virgil A.S.H. Dalm, Harm de Wit, and Hemmo A. Drexhage
Department of Immunology, Erasmus Medical Center, Rotterdam, the Netherlands

Address for correspondence: Virgil A.S.H. Dalm, Department of Immunology, Erasmus MC, Room D-425, s'-Gravendijkwal 230, 3015 CE, Rotterdam, the Netherlands. v.dalm@erasmusmc.nl

Chronic purulent rhinosinusitis (CPR) is an inflammatory condition of unknown origin. Although various medical and surgical treatment modalities are available, 5–10% of patients remain refractory. Immune deficiency is one of the underlying risk factors for this disease. Earlier studies demonstrated disturbances in cell-mediated immunity and defects in monocyte chemotaxis in CPR. Treatment with the thymic hormone preparation thymostimulin led to significant clinical improvement in patients and *in vitro* restoration of monocyte chemotaxis. Unfortunately, thymostimulin became unavailable, which has led to recent interest in the immunomodulatory effects of the thymic peptide thymosin α1, which has demonstrated some benefit for CPR. Our current *in vitro* work focuses on the potential effects of thymosin α1 on monocyte function and gene expression profiles in order to understand its effects and mechanisms of action. Future clinical studies will evaluate the potential significance of thymosin α1 in treatment of CPR patients.

Keywords: thymosin α1; rhinosinusitis; immune deficiency; monocyte; polarization

Chronic purulent rhinosinusitis

Chronic purulent rhinosinusitis (CPR) is known as an inflammatory disorder of the nose and paranasal sinuses of unknown cause defined on the basis of characteristic symptoms (≥ two of the following: nasal congestion, anterior or posterior nasal drainage, facial pain/pressure, and reduced or absent sense of smell), duration (> 12 weeks), and objective evidence of sinus disease by means of direct visualization or imaging studies.[1]

CPR accounts for substantial health care expenditures in terms of lost work days, missed school days, office visits, and antibiotic prescriptions filled.[2] Currently available medical treatment options for CPR include nasal saline,[3] intranasal steroids,[4] and systemic antibiotics.[5] For patients who are refractory to maximal medical therapy endoscopic sinus surgery has been shown to have both short- and long-term benefits in most studies.[6] A significant percentage of patients, around 5–10%, continue to have chronic sinus disease in the face of maximal medical treatment and appropriate surgical intervention.[1]

Different factors have been suggested to predispose to the development of CPR, including asthma and polyps.[7–8] Other factors include irritants in air pollution, such as sulfur dioxide,[9] ozone,[10] and formaldehyde,[11] which all can adversely affect mucociliary clearance. Active cigarette smoking is associated with an increased risk of sinus disease, while there is still debate whether second hand smoking is a true risk factor.[12,13] Although anatomic variants were thought to be involved, there is currently little evidence that these indeed play a role in development of CPR.[14–16] Finally, certain immune deficiency diseases are associated with increased susceptibility to upper respiratory tract infections.[17] In previous studies, an unexpected high incidence of immune dysfunction was found in patients with recurrent sinusitis.[18] Current research in our lab focuses on defects in the innate immune system, more specifically, in human monocytes, that play a role in the pathogenesis of CPR. We also investigate the potential therapeutic effects of the immunomodulatory peptide thymosin α1 in CPR.

doi: 10.1111/j.1749-6632.2012.06712.x

Immune deficiencies in chronic and recurrent sinusitis

The predominant clinical characteristics of immune deficiency diseases are increased frequency and severity of infections. Primary immunodeficiencies (PID) comprise a group of heterogeneous, inherited disorders of the human immune system, including humoral immune deficiencies (characterized by disorders in B cell function), the severe combined immunodeficiencies (disturbances in both B and T cell function), and disorders resulting from defects in phagocyte function or complement activity.[17] In PID, primary antibody deficiencies form the largest group and are characterized by a marked reduction or absence of serum immunoglobulins (Ig) or IgG subclasses and poor response to vaccination.[17,19] Patients with an antibody deficiency can present in early childhood or in adulthood with increased susceptibility to infections typically involving the upper and lower respiratory tracts (otitis, sinusitis, and pneumonia).[20] Common infectious agents are encapsulated bacteria, such as *Streptococcus pneumonia* and *Haemophilus influenzae*.[17]

Chronic or recurrent sinusitis in isolation is rarely associated with a severe antibody deficiency state and may be seen in the less severe immunodeficiency states, such as IgG subclass deficiency, selective antibody deficiency (characterized by a deficient specific antibody response to polysaccharide antigens with normal Ig and IgG subclass concentrations),[21] or deficiency in IgA production. In a previous study, IgA deficiency was found in 6.2% of CPR patients, and total IgG deficiency was found in an unexpectedly high 17.9%.[18]

In a cohort of pediatric patients, 34 of 61 children with refractory sinusitis had abnormal results on immune studies, with decreased IgG_3 levels and poor response to pneumococcal antigen being the most common abnormalities found.[22] In adult patients with chronic rhinosinusitis, Vanlerberghe *et al.* found that IgG_2, IgG_3, or a combined deficiency of major and/or minor IgG subclasses occurred in 22.8% of patients with refractory chronic rhinosinusitis.[23] Recent studies, the first conducted in a tertiary care academic referral center, demonstrated an inadequate response to an unconjugated pneumococcal polysaccharide vaccine in 67%[24] and 11.6%[25] of adult patients with refractory chronic rhinosinusitis.

The complement system is part of the immune system and comprises an important part of the innate (non-specific) immune system, and to some extent it acts as a bridge between the innate and adaptive (specific) immune systems.[26] The complement system, composed of more than 30 components soluble or membrane-bound proteins, plays an important role in the protection against infections.[27] Clinically relevant complement deficiencies are rare; patients with complement deficiencies commonly encounter invasive infections, such as sepsis and meningitis, pneumonia, and upper respiratory tract infections. Common causative agents are encapsulated bacteria.[27] There are few specific data on the significance of complement deficiencies for risk of (chronic) sinusitis; for example, mannose-binding lectin (MBL) deficiency is one of the most prevalent innate immunodeficiencies, but there is little evidence for an increased prevalence of MBL deficiency in patients with chronic rhinosinusitis.[28] On the other hand, deficiency in the complement protein C4 in patients with chronic sinusitis has been described in few reports.[29,30]

T cell–mediated immunity in CPR

Initial studies performed in the 1980s have assessed T cell–mediated immunity to three microorganisms frequently colonizing the human upper respiratory tract, i.e., *H. influenzae*, streptococci, and *Candida albicans* in patients with CPR.[31–32] T cell–mediated immunity in these patients was studied because the somatic antigen of *H. influenzae* was found to be a potent stimulator of the T cell system[33] and disturbances in cell-mediated immunity, as in DiGeorge syndrome, were accompanied by *H. influenzae* infections of the respiratory tract.[34] Delayed type hypersensitivity skin test reactivity, regarded as T cell–mediated responses as similar reactions occurred in rats sensitized with T cell transfers,[33] to somatic antigen of *H. influenzae* was disturbed in 75% of patients, whereas in healthy controls this test was found to be normal in 90% of the individuals tested.[31] This study suggested that abnormalities in cell-mediated immunity form a basis for CPR.

Subsequently, in a second study by the same group, in 75 patients with unexplained CPR, T cell–mediated immunity was assessed as well.[32] Two-thirds of the patients tested showed defective delayed-type hypersensitivity responses, whereas a positive response was found in over 90% of healthy subjects. Moreover, in this study the production of the macrophage migration inhibitory factor (MIF) was measured. MIF was described in 1966[35] and is a pivotal regulator of innate immunity, being a proinflammatory cytokine.[36] *In vitro* measurement of MIF production was found to be the best correlate of delayed-type hypersensitivity reaction *in vivo*.[37] In the study, two-thirds of patients showed defective MIF production.[32] A study in 2001 by a different group confirmed the findings described in the previous studies with respect to cellular immune defects in patients with chronic sinusitis.[18] In 40% of patients, delayed-type hypersensitivity skin test reactivity was disturbed. *In vitro* studies revealed that 26.3% of patients showed abnormally decreased responses to T cell mitogens such as concavalin A, phytohemagglutinins, and anti-CD3. Of the patients, 54.8% showed abnormal proliferation response to the recall antigens mumps and tetanus, and 11.3% had decreased response to alloantigen.[18] The results of these three studies clearly demonstrated the presence of defects in T cell reactivity toward microbial antigens in a high number of patients suffering from CPR.

Defects in monocyte polarization in CPR

T cell abnormalities are often accompanied by an impaired function of monocytes, as has been documented for atopic dermatitis, in which monocyte chemotaxis was found to be frequently depressed.[38–39] In patients with AIDS[40] and immune deficiencies accompanying various types of malignancies,[41–43] defects in monocyte chemotaxis have been described. In malignancies, the presence in serum of factors capable of inhibiting monocyte chemotactic responsiveness was reported in the 1980s.[43–44] These factors appeared to share structural homology to p15E, which is a feline and murine retroviral transmembrane protein;[44] p15E-like endogenous retroviral factors are produced by immune cells such as lymphocytes and monocytes, suggesting a role in immune regulation.[45]

Following the studies on defective T cell–mediated immunity in CPR, the defects in chemotactic responsiveness of monocytes in patients with CPR have been studied in more detail. The chemotactic responsiveness was determined by using a

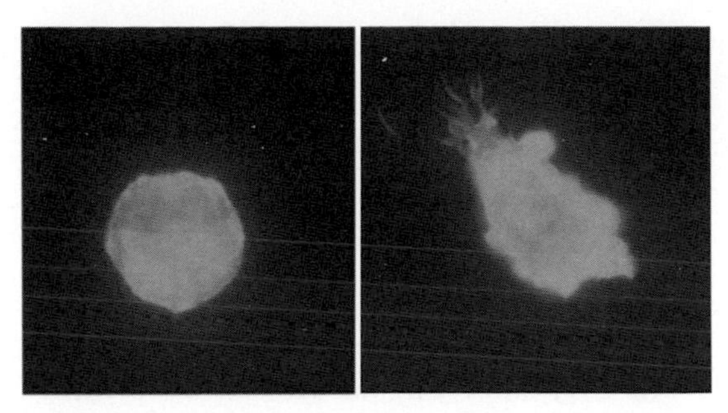

Figure 1. A monocyte in resting state (left) and a polarizing monocyte (right).

polarization assay, as the polarization of human monocytes toward chemoattractants is an early event preceding their chemotactic response.[46] Polarization is the change of shape from a round to elongated or triangular form, with sometimes large cytoplasmatic protrusions. Figure 1 shows an image of a resting and polarized monocyte. The ability to undergo impressive morphological changes is required for monocyte passage through small capillary vessels, migration from blood to inflamed tissue, ingestion of foreign particles, and for traveling along a chemotactic gradient.[47,48]

In an initial study of 40 patients with CPR with previously established defects in T cell–mediated immunity, an impaired monocyte chemotactic responsiveness was found in 60% of patients.[49] The polarization assays were performed using *N*-formyl-methionyl-leucyl-phenylalanine (fMLP), a so-called *N*-formyl peptide, which is either derived from bacterial proteins or from endogenous mitochondrial proteins[50] and binds to the G-protein coupled chemoattractant receptor formyl peptide receptor (FPR).[51] In these patients with hampered chemotactic responsiveness, p15E-related inhibitory factors were found in serum as well, and it was hypothesized that these factors could interfere with the receptors for fMLP. In a subsequent study by another group in patients with chronic rhinosinusitis with or without nasal polyps, defects in cellular immunity associated with the presence of p15E-related proteins were found in approximately 90% of patients.[52] These findings suggested that chronic inflammation of the upper airway epithelium in general was associated with defects in cellular immunity.[52]

Disturbances in monocyte chemotactic responsiveness: a new target for therapy in CPR

As cell-mediated immunity and disturbances in monocyte chemotactic responsiveness were found to be of importance in the pathophysiology of CPR, in the early 1990s novel treatment strategies targeting these disturbances were studied in CPR patients.

The first study on this subject was published in 1990.[53] It was known that thymic hormones had strong effects on the functioning of the cell-mediated immune system. Thymic hormones induce a terminal differentiation in T cells and are able to stimulate T lymphocytes to produce cytokines. Therefore, in this first study the effects of the bovine thymic hormone preparation thymostimulin (TP-1®, Serono), a standardized low-molecular protein fraction containing thymic humoral factor and thymosin α1, were studied in a double-blinded crossover trial in 20 patients with CPR and proven impairments in cell-mediated immunity. In the study, thymostimulin was injected intramuscularly. Eighteen out of 20 patients completed the trial and showed clinical improvement. On inspection, nasal mucosa was absent of mucopurulent secretion in 13 patients, and positive nasal bacterial culture rates decreased from 14 out of 16 to 5 out of 15. Clinical improvement was correlated with improved chemotactic responsiveness, as assessed in the monocyte polarization assay, and suppressive p15E-like factors in serum decreased during treatment. Thus, this study reported for the first time that patients suffering from chronic purulent rhinosinusitis could be successfully treated with thymic hormone preparations, restoring monocyte polarization, and neutralizing the activity of p15E-like factors.[53] However, TP-1 was taken off the market and further clinical studies with TP-1 have not been performed.

Into the 21st century: thymosin α1 for treatment of CPR?

In recent years interest in the potential beneficial effects of thymic hormones and thymic hormone preparations on immune cell function has increased. In 1966, U.S. investigators extracted a so-called lymphocytopoietic factor from calf thymus. This factor stimulated lymphocyte proliferation *in vitro* and in animal models, was initially believed to be a single polypeptide, and was named "thymosin."[54] A further five-step purification led to thymosin fraction 5, identified as a mixture of 30 to 40 small polypeptide components.[55] Thymosin α1 is a 28 amino acid peptide isolated from thymosin fraction 5.[55] The most relevant immunomodulatory properties of thymosin α1 include increase of cytokine production and natural killer cell activity, enhancement of interleukin-2 receptor expression, induction of phenotypic T cell marker expression, increase of major histocompatibility complex class I expression, and enhancement of resistance to viruses, mycobacteria, and fungi.[56] Moreover, it was found that thymosin α1 enhances dendritic cell differentiation from human peripheral blood monocytes.[57] Today, thymosin α1 is produced synthetically and is approved mainly in Asia and South America for the treatment of chronic hepatitis B and C and as a vaccine enhancer, and in few countries in Southeast Asia for the treatment of cancer.[58] In a recent double-blind randomized control study in patients with severe acute pancreatitis it was found that treatment with thymosin α1 improved cell-mediated immunity and reduced infection rate.[59]

Current and future perspectives

As a significant percentage of CPR patients are refractory to the currently available treatment options, novel therapies are needed. In the past, beneficial effects of thymic peptides on clinical measures in CPR have been described. With thymosin α1 now commercially available, it would be of great interest to further explore the potential significance of this peptide in CPR.

In ongoing studies in our lab, we are investigating whether thymosin α1 could be a potential novel therapeutic option in CPR patients with estimated disturbances in monocyte chemotactic responsiveness. Monocytes from CPR patients and healthy controls are isolated and monocyte polarization assays are performed in order to select the patients with diminished monocyte polarization *in vitro*. Improvement of polarization is studied upon stimulation of the monocytes with thymosin α1. Some preliminary results of ongoing research are outlined below (Fig. 2).

Healthy controls (12 individuals) have thus far been evaluated. In these subjects a mean polarization of 58% was found with a standard deviation (SD) of 12.5%. We found no differences between

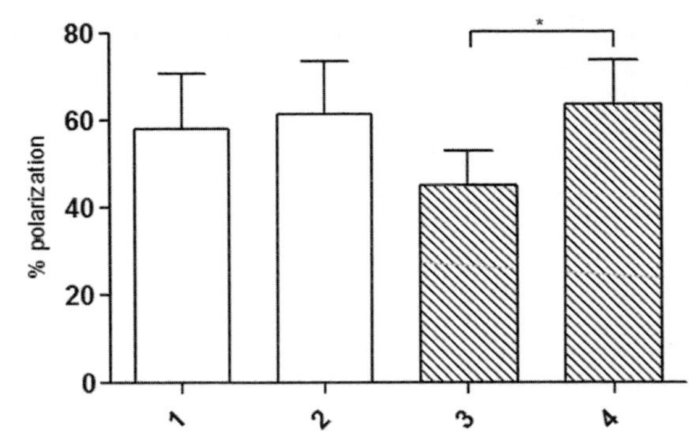

Figure 2. Monocyte chemotaxis was tested using the modified Ciancolo and Snyderman monocyte polarization assay. Monocyte polarization percentage is represented on the vertical axis; the horizontal axis represents four different study groups as follows: (1) healthy controls ($n = 12$), (2) healthy controls + thymosin α1, (3) CPR patients ($n = 4$), and (4) CPR patients + thymosin α1. Four CPR patients were found to have defective polarization (bar 3) and *in vivo* treatment with thymosin α1 led to normalization of the polarization of these monocytes (bar 4). $^*P < 0.05$. Data presented here are work in progress.

male and female subjects; with females showing a mean of 59%, SD 14% ($n = 8$) and males a mean of 57%, SD 11% ($n = 4$). Out of 16 patients selected from the outpatient clinics of the department of immunology, a mean polarization of 54%, SD 11.5% was found. Monocytes from four patients were found to have diminished polarization, with a mean of 45%, SD 8%, compared with healthy controls, although statistical significance was not reached ($P = 0.078$). Monocytes of both healthy controls and patients were also incubated with thymosin α1 and we found that thymosin α1 only significantly improved the polarization of monocytes in CPR patients (mean 45%, SD 8% versus mean 64% SD 10%, $P = 0.029$). In either healthy controls or CPR patients without diminished polarization, no significant improvement was shown. Figure 2 shows the results of the polarization assay in 12 healthy controls and 4 patients with a diminished monocyte polarization before and after treatment with thymosin α1.

These preliminary results are in line with previous studies showing defective monocyte polarization in patients with CPR and improvement upon stimulation with thymosin α1. Moreover, these findings provide a basis for further studies on the clinical effects of thymosin α1 in CPR patients. In ongoing studies, with whole genome microarray analysis, followed by Q-PCR on whole-genome–detected discriminating genes, we will evaluate the gene expression profiles of monocytes obtained from patients compared to healthy controls in order to gain more insight in the potential deficits in monocyte populations in patients suffering from CPR. Particular focus will be on the difference between patients with and without monocyte polarization defects in order to identify molecules testable in Q-PCR that could replace and/or complete the functional polarization assay. Moreover, the effects of thymosin α1 on gene expression profiles in these monocytes will be investigated as well.

Finally, a clinical study will have to be performed to determine the potential clinical significance of thymosin α1 in patients with CPR. Based on previous studies and our recent findings on the effects of thymosin α1 on monocyte polarization *in vitro*, we feel that thymosin α1 may be a future therapeutic option for those CPR patients who are refractory to the current available treatment regimens.

Conflicts of interest

The authors declare no conflicts of interest.

References

1. Benninger, M.S. *et al.* 2003. Adult chronic rhinosinusitis: definitions, diagnosis, epidemiology, and pathophysiology. *Otolaryngol. Head. Neck. Surg.* **129:** S1–S32.
2. Bhattacharyya, N. 2009. Contemporary assessment of the disease burden of sinusitis. *Am. J. Rhinol. Allergy* **23:** 392–395.
3. Hauptman, G. & M.W. Ryan. 2007. The effect of saline solutions on nasal patency and mucociliary clearance in rhinosinusitis patients. *Otolaryngol. Head. Neck. Surg.* **137:** 815–821.
4. Parikh, A. *et al.* 2001. Topical corticosteroids in chronic rhinosinusitis: a randomized, double-blind, placebo-controlled trial using fluticasone propionate aqueous nasal spray. *Rhinology* **39:** 75–79.
5. Tichenor, W.S. *et al.* 2008. Nasal and sinus endoscopy for medical management of resistant rhinosinusitis, including postsurgical patients. *J. Allergy Clin. Immunol.* **121:** 917–927 e2.
6. Lund, V.J. *et al.* 1995. The treatment of chronic sinusitis: a controversial issue. *Int. J. Pediatr. Otorhinolaryngol.* **32** (Suppl): S21–S35.
7. Bachert, C. *et al.* 2010. Rhinosinusitis and asthma: a link for asthma severity. *Curr. Allergy Asthma. Rep.* **10:** 194–201.
8. Pearlman, A.N. *et al.* 2009. Relationships between severity of chronic rhinosinusitis and nasal polyposis, asthma, and atopy. *Am. J. Rhinol. Allergy* **23:** 145–148.
9. Kienast, K. *et al.* 1996. Combined exposures of human ciliated cells to different concentrations of sulfur dioxide and nitrogen dioxide. *Eur. J. Med. Res.* **1:** 533–536.

10. Christian, D.L. *et al.* 1998. Ozone-induced inflammation is attenuated with multiday exposure. *Am. J. Respir. Crit. Care. Med.* **158:** 532–537.
11. Schafer, D. *et al.* 1999. In vivo and in vitro effect of ozone and formaldehyde on human nasal mucociliary transport system. *Rhinology* **37:** 56–60.
12. Lieu, J.E. & A.R. Feinstein. 2000. Confirmations and surprises in the association of tobacco use with sinusitis. *Arch. Otolaryngol. Head Neck Surg.* **126:** 940–946.
13. Reh, D.D. *et al.* 2009. Secondhand tobacco smoke exposure and chronic rhinosinusitis: a population-based case-control study. *Am. J. Rhinol. Allergy* **23:** 562–567.
14. Bolger, W.E., C.A. Butzin & D.S. Parsons. 1991. Paranasal sinus bony anatomic variations and mucosal abnormalities: CT analysis for endoscopic sinus surgery. *Laryngoscope* **101:** 56–64.
15. Jones, N.S., A. Strobl & I. Holland. 1997. A study of the CT findings in 100 patients with rhinosinusitis and 100 controls. *Clin. Otolaryngol. Allied Sci.* **22:** 47–51.
16. Lusk, R.P., B. McAlister & A. el Fouley. 1996. Anatomic variation in pediatric chronic sinusitis: a CT study. *Otolaryngol. Clin. North Am.* **29:** 75–91.
17. Notarangelo, L.D. 2010. Primary immunodeficiencies. *J. Allergy Clin. Immunol.* **125:** S182–S194.
18. Chee, L. *et al.* 2001. Immune dysfunction in refractory sinusitis in a tertiary care setting. *Laryngoscope* **111:** 233–235.
19. van der Burg, M. *et al.* 2012. New frontiers of primary antibody deficiencies. *Cell Mol. Life Sci.* **69:** 59–73.
20. Oksenhendler, E. *et al.* 2008. Infections in 252 patients with common variable immunodeficiency. *Clin. Infect. Dis.* **46:** 1547–1554.
21. Ambrosino, D.M. *et al.* 1987. An immunodeficiency characterized by impaired antibody responses to polysaccharides. *N. Engl. J. Med.* **316:** 790–793.
22. Shapiro, G.G. *et al.* 1991. Immunologic defects in patients with refractory sinusitis. *Pediatrics* **87:** 311–316.
23. Vanlerberghe, L., S. Joniau & M. Jorissen. 2006. The prevalence of humoral immunodeficiency in refractory rhinosinusitis: a retrospective analysis. *B-ENT.* **2:** 161–166.
24. Alqudah, M., S.M. Graham & Z.K. Ballas. 2010. High prevalence of humoral immunodeficiency patients with refractory chronic rhinosinusitis. *Am. J. Rhinol. Allergy* **24:** 409–412.
25. Carr, T.F. *et al.* 2011. Characterization of specific antibody deficiency in adults with medically refractory chronic rhinosinusitis. *Am. J. Rhinol. Allergy* **25:** 241–244.
26. Dunkelberger, J.R. & W.C. Song. 2010. Complement and its role in innate and adaptive immune responses. *Cell Res.* **20:** 34–50.
27. Skattum, L. *et al.* 2011. Complement deficiency states and associated infections. *Mol. Immunol.* **48:** 1643–1655.
28. Dahl, M. *et al.* 2004. A population-based study of morbidity and mortality in mannose-binding lectin deficiency. *J. Exp. Med.* **199:** 1391–1399.
29. Ogunleye, A.O. & O.G. Arinola. 2001. Immunoglobulin classes, complement factors and circulating immune complexes in chronic sinusitis patients. *Afr. J. Med. Med. Sci.* **30:** 309–312.
30. Seppanen, M. *et al.* 2006. Immunoglobulins and complement factor C4 in adult rhinosinusitis. *Clin. Exp. Immunol.* **145:** 219–227.
31. Drexhage, H.A. *et al.* 1983. Abnormalities in cell-mediated immune functions to Haemophilus influenzae chronic purulent infections of the upper respiratory tract. *Clin. Immunol. Immunopathol.* **28:** 218–228.
32. van de Plassche-Boers, E.M. *et al.* 1986. Parameters of T cell mediated immunity to commensal micro-organisms in patients with chronic purulent rhinosinusitis: a comparison between delayed type hypersensitivity skin test, lymphocyte transformation test and macrophage migration inhibition factor assay. *Clin. Exp. Immunol.* **66:** 516–524.
33. Drexhage, H.A. & J. Oort. 1977. Skin test reactivity to H. influenzae antigens as an outcome of the antigen structure and the balance between humoral and cell-mediated immunity in rats. *Clin. Exp. Immunol.* **28:** 280–288.
34. Kornfeld, S.J. *et al.* 2000. DiGeorge anomaly: a comparative study of the clinical and immunologic characteristics of patients positive and negative by fluorescence in situ hybridization. *J. Allergy Clin. Immunol.* **105:** 983–987.
35. David, J.R. 1966. Delayed hypersensitivity in vitro: its mediation by cell-free substances formed by lymphoid cell-antigen interaction. *Proc. Natl. Acad. Sci. USA* **56:** 72–77.
36. Kerschbaumer, R.J. *et al.* 2012. Neutralization of macrophage migration inhibitory factor (MIF) by fully human antibodies correlates with their specificity for the beta-sheet structure of MIF. *J. Biol. Chem.* **287:** 7446–7455.
37. Thor, D.E. *et al.* 1968. Cell migration inhibition factor released by antigen from human peripheral lymphocytes. *Nature* **219:** 755–757.
38. Furukawa, C.T. & L.C. Altman. 1978. Defective monocyte and polymorphonuclear leukocyte chemotaxis in atopic disease. *J. Allergy Clin. Immunol.* **61:** 288–293.
39. Snyderman, R., E. Rogers & R.H. Buckley. 1977. Abnormalities of leukotaxis in atopic dermatitis. *J. Allergy Clin. Immunol.* **60:** 121–126.
40. Poli, G. *et al.* 1985. Monocyte function in intravenous drug abusers with lymphadenopathy syndrome and in patients with acquired immunodeficiency syndrome: selective impairment of chemotaxis. *Clin. Exp. Immunol.* **62:** 136–142.
41. Currie, G.A. & D.W. Hedley. 1977. Monocytes and macrophages in malignant melanoma. I. Peripheral blood macrophage precursors. *Br. J. Cancer* **36:** 1–6.
42. Snyderman, R. *et al.* 1978. Abnormal monocyte chemotaxis in patients with breast cancer: evidence for a tumor-mediated effect. *J. Natl. Cancer Inst.* **60:** 737–740.
43. Tan, I.B. *et al.* 1986. Defective monocyte chemotaxis in patients with head and neck cancer. Restoration after treatment. *Arch. Otolaryngol. Head Neck Surg.* **112:** 541–544.
44. Cianciolo, G. *et al.* 1981. Inhibitors of monocyte responses to chemotaxins are present in human cancerous effusions and react with monoclonal antibodies to the P15(E) structural protein of retroviruses. *J. Clin. Invest.* **68:** 831–844.
45. Tan, I.B. *et al.* 1987. Immunohistochemical detection of retroviral-P15E-related material in carcinomas of the head and neck. *Otolaryngol. Head Neck Surg.* **96:** 251–255.
46. Cianciolo, G.J. & R. Snyderman. 1981. Monocyte responsiveness to chemotactic stimuli is a property of a subpopulation

of cells that can respond to multiple chemoattractants. *J. Clin. Invest.* **67:** 60–68.
47. Prossnitz, E.R. & R.D. Ye. 1997. The N-formyl peptide receptor: a model for the study of chemoattractant receptor structure and function. *Pharmacol. Ther.* **74:** 73–102.
48. Zaffran, Y. *et al.* 1993. Role of calcium in the shape control of human granulocytes. *Blood Cells* **19:** 115–129; discussion 129–131.
49. van de Plassche-Boers, E.M. *et al.* 1988. Abnormal monocyte chemotaxis in patients with chronic purulent rhinosinusitis: an effect of retroviral p15E-related factors in serum. *Clin. Exp. Immunol.* **73:** 348–354.
50. Marasco, W.A. *et al.* 1984. Purification and identification of formyl-methionyl-leucyl-phenylalanine as the major peptide neutrophil chemotactic factor produced by Escherichia coli. *J. Biol. Chem.* **259:** 5430–5439.
51. Le, Y., P.M. Murphy & J.M. Wang. 2002. Formyl-peptide receptors revisited. *Trends Immunol.* **23:** 541–548.
52. Scheeren, R.A. *et al.* 1993. Defects in cellular immunity in chronic upper airway infections are associated with immunosuppressive retroviral p15E-like proteins. *Arch. Otolaryngol. Head Neck Surg.* **119:** 439–443.
53. Tas, M., J.A. Leezenberg & H.A. Drexhage. 1990. Beneficial effects of the thymic hormone preparation thymostimulin in patients with defects in cell-mediated immunity and chronic purulent rhinosinusitis. A double-blind cross-over trial on improvements in monocyte polarization and clinical effects. *Clin. Exp. Immunol.* **80:** 304–313.
54. Goldstein, A.L., F.D. Slater & A. White. 1966. Preparation, assay, and partial purification of a thymic lymphocytopoietic factor (thymosin). *Proc. Natl. Acad. Sci. USA* **56:** 1010–1017.
55. Goldstein, A.L. *et al.* 1977. Thymosin alpha1: isolation and sequence analysis of an immunologically active thymic polypeptide. *Proc. Natl. Acad. Sci. USA* **74:** 725–729.
56. Garaci, E. 2007. Thymosin alpha1: a historical overview. *Ann. N.Y. Acad. Sci.* **1112:** 14–20.
57. Yao, Q. *et al.* 2007. Thymosin-alpha1 modulates dendritic cell differentiation and functional maturation from human peripheral blood CD14+ monocytes. *Immunol. Lett.* **110:** 110–120.
58. Billich, A. 2002. Thymosin alpha1. SciClone Pharmaceuticals. *Curr. Opin. Invest. Drugs* **3:** 698–707.
59. Wang, X. *et al.* 2011. Thymosin alpha 1 is associated with improved cellular immunity and reduced infection rate in severe acute pancreatitis patients in a double-blind randomized control study. *Inflammation* **34:** 198–202.

Ann. N.Y. Acad. Sci. ISSN 0077-8923

ANNALS OF THE NEW YORK ACADEMY OF SCIENCES
Issue: *Thymosins in Health and Disease*

Thymosin α1 in melanoma: from the clinical trial setting to the daily practice and beyond

Riccardo Danielli, Ester Fonsatti, Luana Calabrò, Anna Maria Di Giacomo, and Michele Maio
Department of Oncology, Division of Medical Oncology and Immunotherapy, University Hospital of Siena, Istituto Toscano Tumori, Siena, Italy

Address for correspondence: Michele Maio, Division of Medical Oncology and Immunotherapy, Department of Oncology, University Hospital of Siena, Viale Mario Bracci 16, 53100 Siena, Italy. mmaio@cro.it

Thymosin α1 (Tα1) is an immunomodulatory peptide released by the thymus gland in mammals. It was first described in 1977 as a potential agent for the treatment of immune deficiencies and cancer. Among solid tumors, a number of clinical trials have investigated the activity of Tα1 in melanoma. In particular, a large randomized phase II trial that evaluated the safety and efficacy of combining Tα1 with dacarbazine and interferon alpha in metastatic melanoma patients provided the rationale for further clinical applications. The main findings emerging from clinical trials and that support the therapeutic use of Tα1 in human melanoma are summarized and discussed.

Keywords: thymosin α1; melanoma; immunotherapy; dacarbazine

Introduction

The incidence of melanoma is increasing worldwide and the prognosis of metastatic patients remains dismal.[1] A recent meta-analysis of 42 phase II trials reported a median survival of only 6.2 months for advanced melanoma patients, with a one-year survival rate of 25.5%, regardless of treatment regimens.[2] Based on this evidence, the immunologic features of melanoma, and the lack of effective therapies for the metastatic disease, major expectations rely on immunotherapy as a potentially effective treatment for melanoma. Along this line, different immunotherapeutic compounds are currently providing initial proof of clinical success, generally being used as single agents. Among these are cancer vaccines,[3] immunomodulating monoclonal antibodies (mAbs) such as the anti-cytotoxic T lymphocyte antigen (CTLA)-4 and the anti-PD-1 mAbs,[4] and the BRAF kinase inhibitors.[5] In particular, it has been recently shown that the anti-CTLA-4 mAb ipilimumab and the BRAF inhibitor vemurafenib, administered alone, significantly improve the overall survival (OS) of metastatic melanoma patients in phase III trials.[6,7] Despite these highly encouraging results when used alone, immunotherapy in combination with chemotherapy has also shown to improve the OS in melanoma patients. The first demonstration derived from a phase III study that evaluated the clinical efficacy of ipilimumab plus dacarbazine (DTIC) versus DTIC plus placebo in previously untreated metastatic melanoma patients.[8] Supporting this initial evidence to the determination that results from a recently concluded phase II trial, combining ipilimumab and fotemustine, have shown significant improvement in disease control in patients affected by metastatic cutaneous melanoma, with or without brain metastases.[9] Based on these initial and highly interesting results, a major future challenge will be to design novel therapeutic strategies to synergistically combine the activity of immunomodulatory agents with other therapeutic agents, including cytotoxic chemotherapy.

A large body of available preclinical *in vitro* and *in vivo* evidence points to thymosin alpha 1 (Tα1) as a useful immunomodulatory peptide, with significant therapeutic potential in metastatic melanoma in the absence of clinically meaningful toxicity.[10] Consistently, the results emerging from initial trials provide support of the ability of Tα1 to improve the clinical outcome of advanced melanoma patients through

doi: 10.1111/j.1749-6632.2012.06757.x

the activation of the immune system.[10,11] Together, these studies provide the rationale for combining Tα1 with new therapeutic agents currently approved or in clinical development for the treatment of advanced melanoma patients.

Tα1 mechanism of action

Tα1, a 28 amino acid peptide of ~3.1 kDa, is endogenously produced by the thymus gland by the cleavage of its precursor pro-Tα1.[12] Although the fine immunologic mechanism(s) of action of Tα1 have not fully been elucidated, experimental evidence points to its strong immunomodulatory properties. In particular, it was reported that Tα1 enhances T cell–mediated immune responses by several mechanisms, including increased T cell production (i.e., $CD4^+$, $CD8^+$, and $CD3^+$ cells), stimulation of T cell differentiation and/or maturation, reduction of T cell apoptosis, and restoration of T cell–mediated antibody production.[12] Furthermore, it was demonstrated that Tα1 acts on the immune system by modulating the release of proinflammatory cytokines (i.e., interleukin-2 (IL-2), interferon-gamma (IFN-γ)),[12–14] and through the activation of natural killer and dendritic cells.[12] In addition, Tα1 was also demonstrated to have direct effects on cancer cells by increasing the levels of expression of different tumor antigens and of components of the major histocompatibility complex class I, as well as by reducing cancer cell growth.[12,13,15]

Together, these experimental findings bear relevance for cancer immunotherapy and suggest that Tα1 can activate innate and adaptive immune responses and modulate the immunophenotype of cancer cells, improving their immunogenicity and their recognition by the immune system.

Tα1 preclinical evidence of antitumor activity

The effects of Tα1 administered alone or in combination with chemotherapy and different cytokines such as IFN-α and IL-2, or a natural Type 1-cytokine mixture, have been investigated in both normal and/or immunosuppressed animals. In 1990, Garaci *et al.*, demonstrated the possibility of completely eradicating Lewis lung carcinoma (3LL) in mice using Tα1, followed by the administration of murine IFN-α/β after cyclophosphamide.[12] This combination treatment proved to be highly effective in curing established tumors, thus demonstrating the ability of Tα1 to synergize with cytokines.[12] In contrast to the combination of therapies, single treatments were not effective. Although the mechanism underlying the activity of Tα1 administered with cyclophosphamide was not fully understood, it was suggested that chemotherapy could directly reduce the tumor mass, whereas Tα1 plus IFN would potentiate the immune system compromised by chemotherapy.[12] Initially, this therapeutic combination was evaluated aiming to synergize the effect of Tα1 and immunoactive molecules at low doses, thus reducing the overall toxicity of treatment. We investigated, based on the results obtained, combining Tα1 plus a cytokine after chemotherapy. Thus, encouraging results were also reported in different experimental models, including friend erythroleukemia and B16 melanoma in mice and DHD/K12 colorectal carcinoma liver metastases in rats.[12]

These *in vivo* findings prompted the evaluation of Tα1 for the treatment of human cancer.[10] The results that emerged from pilot clinical trials in human hepatocellular carcinoma, non-small-cell lung cancer, and melanoma suggested that Tα1 can improve the outcome of cancer patients and demonstrated that Tα1 has a very favorable toxicity profile.[10]

Tα1 clinical activity in melanoma

Because melanoma immunobiology has been extensively investigated, it represents a useful tumor model for the development of new immunotherapeutic and targeted treatments. In addition, it is likely that therapeutic insights generated in melanoma can be transferred to other solid malignancies. With regard to the clinical use of Tα1 in melanoma, we will focus on the results of a large randomized phase II study and on the experience of a single institution within a Tα1 compassionate use program.

Thus, so far the results from one clinical trial using Tα1 have been reported in patients with metastatic cutaneous melanoma.[11] This large, randomized, phase II study was conducted at 64 European centers between 2002 and 2006 to investigate the efficacy of Tα1 administered in combination with DTIC or with DTIC + IFN-α, versus only DTIC + IFN-α, in 488 previously untreated patients with cutaneous metastatic melanoma.[11] The study was designed to evaluate the ability of Tα1 to potentiate the therapeutic efficacy of DTIC.

Patients with confirmed (either histologically or cytologically) stage IV melanoma (American Joint Cancer Committee),[16] with unresectable metastases and one or more measurable lesions, and with an Eastern Cooperative Oncology Group performance status of 0 to 1, were randomly assigned to five treatment groups: DTIC + IFN-α and 1.6 mg of Tα1; DTIC + IFN-α and 3.2 mg of Tα1; DTIC + IFN-α and 6.4 mg of Tα1; DTIC + 3.2 mg of Tα1; and DTIC + IFN-α. Treatment was continued until disease progression, development of any serious adverse event (AE), or withdrawal of consent.[11] In summary, the results demonstrated that the clinical benefit rate (CBR), defined as the proportion of patients with a complete response, partial response, or stable disease, was significantly higher in patients who received Tα1 + DTIC than in those who received control therapy. There was also a trend for improved progression-free survival (hazard ratio (HR): 0.80; 95% confidence interval (CI): 0.63–1.01; $P = 0.06$) and OS (median: 9.4 vs. 6.6 months, respectively; HR: 0.80; 95% CI: 0.63–1.02; $P = 0.08$) in patients who received Tα1 (all groups combined) compared with those who received the control treatment.[11] The data that emerged from this study suggested that the addition of Tα1 to standard DTIC treatment resulted in a reduction in the risk of mortality and disease progression in patients with metastatic malignant melanoma, and pointed to a poor effect of IFN-α in the combination. This latter finding is likely because of the limited therapeutic activity of IFN-α even in the adjuvant setting in stage III melanoma patients.[17] Of note, the therapeutic efficacy of Tα1 in combination with DTIC was associated with very limited toxicity for treated patients,[11] who also experienced long-term benefit from Tα1 (Fig. 1, preliminary work in progress). These encouraging results of Tα1 in combination with DTIC provide the basis to test the association in a phase III trial in melanoma. The primary end point of this study was overall response rate evaluated as per Response Evaluation Criteria in Solid Tumors (RECIST) 1.0 criteria.[11]

Given that the kinetics of activity of immunotherapeutic agents are quite different than they are for cytotoxic agents, new criteria to evaluate the clinical response to immunotherapy, other than redefining trial endpoints, were published by Wolchock *et al.*,[18] who pointed out that standard assessment criteria, such as RECIST or those indicated by the World Health Organization (WHO), conventionally applied to cytotoxic agents, do not adequately capture some patterns of response observed in the course of immunotherapy; stemming from these considerations, immune-related response criteria (irRC) were developed to measure primary and secondary endpoints in immunotherapy clinical trials.[18] The endpoints and assessment criteria used in the large randomized study with Tα1 + DTIC described above likely underestimated the therapeutic efficacy of Tα1 in melanoma patients, as irRC were not used.

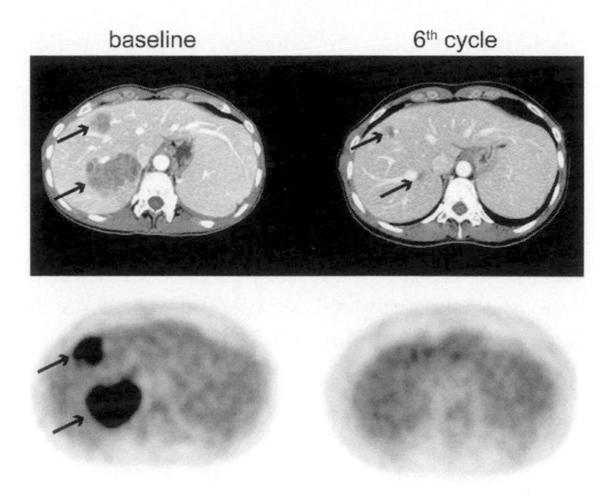

Figure 1. Long-term objective response observed within the Tα1 phase II study. Positron emission tomography (PET) and computed tomography (CT) scans in a 39-year-old female affected by metastatic cutaneous melanoma in the liver treated within the Tα1 phase II study. Multiple hepatic metastases were identified by CT and PET scans at baseline (upper and lower left images, respectively); after six cycles of treatment, only fibrotic nodules, without metabolic activity, were identified by CT and PET scans (upper and lower right images, respectively). The patient is still alive and in complete clinical remission five years from the first dose of treatment.

A single-institution experience with Tα1 + DTIC in 31 advanced-stage malignant melanoma patients was recently conducted within a compassionate use program at our institution; the results of this experience have been collected and will be presented elsewhere.[19] However, the observed CBR of 41% anticipates that the combination of Tα1 + DTIC can be effective in heavily pretreated metastatic melanoma patients without significant toxicity. These findings, though generated from a small population of patients, further support the promising therapeutic role of Tα1 in combination with DTIC in metastatic melanoma.

The high rate of disease stabilization observed with Tα1 in the course of these studies is typical of immunotherapy, where reductions in tumor volume occur slowly and steadily with ongoing treatment.[18] In addition, a sizeable proportion of treated patients experienced a long-term stabilization or complete remission of disease that lasted even for several years (Fig. 1). Notably, this type of response was clearly identified in several studies, as well as in an expanded access program of CTLA-4 mAb ipilimumab treatment[20] that has recently received approval by the Food and Drug Administration (FDA) in the United States and by the European Medicinal Agency in Europe. The results obtained in the clinical setting with Tα1 thus far suggest that further investigation into more appropriately designed clinical trials is warranted—trials that take into account the time that immunotherapy takes to exploit any clinical potential in full. Thus, based on evidence showing a favorable safety profile even in long-term responders,[11] it can be envisaged that novel combination studies should explore the activity of Tα1 in association with other approved agents, such as ipilimumab and vemurafenib or as maintenance therapy in melanoma patients who experience clinical benefit after treatment with these agents. In addition, because of the pleiotropic immune mechanism(s) of action of Tα1, including the upregulation of T cell–driven immune responses against specific tumor antigens, priming of immune responses and potentiation of antitumor T cell–mediated immune responses through the activation of Toll-like receptor 9 on dendritic cells,[21–23] coupling Tα1 to cancer vaccines should be an additional useful therapeutic strategy to pursue. Tα1 could, in fact, prove helpful in overcoming the limited immunogenicity and the short-lived persistency of adequate immunologic antitumor responses frequently reported as potential causes of failure of therapeutic vaccines.[23]

In summary, the results obtained in the course of available clinical experiences have shown that Tα1 is well tolerated and has clinical potential in different clinical settings in melanoma. Combination trials with immunomodulating antibodies, new target therapies, and cancer vaccines will most likely be the key to further exploit therapeutic potential of Tα1 in the melanoma clinic.

Conflicts of interest

The authors declare no conflicts of interest.

References

1. Lui, P. *et al.* 2007. Treatments for metastatic melanoma: synthesis of evidence from randomized trials. *Cancer Treat. Rev.* **33:** 665–680.
2. Korn, E.L. *et al.* 2008. Meta-Analysis of Phase II Cooperative Group Trials in Metastatic Stage IV Melanoma to Determine Progression-Free and Overall Survival Benchmarks for Future Phase II Trials. *J. Clin. Oncol.* **4:** 527–534.
3. Kirkwood, J.M. *et al.* 2008. Next generation of immunotherapy for melanoma. *J. Clin. Oncol.* **26:** 3445–3455.
4. Sznol, M. 2012. Advances in the treatment of metastatic melanoma: new immunomodulatory agents. *Semin. Oncol.* **39:** 192–203.
5. Kudchadkar, R. *et al.* 2012. Targeting mutant BRAF in melanoma: current status and future development of combination therapy strategies. *Cancer J.* **18:** 124–131.
6. Hodi, F.S. *et al.* 2010. Improved survival with ipilimumab in patients with metastatic melanoma. *N. Engl. J. Med.* **363:** 711–723.
7. Chapman, P. *et al.* 2011. Improved survival with vemurafenib in melanoma with BRAF V600E mutation. *N. Engl. J. Med.* **364:** 2507–2516.
8. Robert, C. *et al.* 2011. Ipilimumab plus dacarbazine for previously untreated metastatic melanoma. *N. Engl. J. Med.* **364:** 2517–2526.
9. Di Giacomo, A.M. *et al.* 2012. A phase II, open label, single-arm study to investigate the combination of ipilimumab and fotemustine in patients with advanced melanoma: the NIBIT-M1 trial. *Lancet Oncol.* **13:** 879–886.
10. Garaci, E. *et al.* 1995. Sequential chemoimmunotherapy for advanced non-small cell lung cancer using cisplatin, etoposide, thymosin-alpha 1 and interferon-alpha 2a. *Eur. J. Cancer.* **31A:** 2403–2405.
11. Maio, M. *et al.* 2010. Large randomized study of thymosin alpha 1, interferon alfa, or both in combination with dacarbazine in patients with metastatic melanoma. *J. Clin. Oncol.* **28:** 1780–1787.
12. Garaci, E. *et al.* 1990. Combination treatment using thymosin alpha one and interferon after cyclophosphamide is able to cure Lewis lung carcinoma in mice. *Cancer Immunol. Immunother.* **32:** 154–160.
13. Rasi, G. *et al.* 1994. Anti-tumor effect of combined treatment with thymosin alpha 1 and interleukin-2 after 5-fluorouracil in liver metastase from colorectal cancer in rats. *Int. J. Cancer.* **57:** 701–705.
14. Giuliani, C. *et al.* 2000. Thymosin-alpha1 regulates MHC class I expression in FRTL-5 cells at transcriptional level. *Eur. J. Immunol.* **30:** 778–786.
15. Sinibaldi Vallebona, P. *et al.* 2002. Thymalfasin up-regulates tumor antigen expression in colorectal cancer cells. *Tumour Biol.* **23:** 45.
16. Balch, C.M. *et al.* 2001. Prognostic factor analysis of 17,600 melanoma patients: validation of the American Joint Committee on Cancer melanoma staging system. *J. Clin. Oncol.* **19:** 3622-3634.
17. Eggermont, A.M. *et al.* 2008. Adjuvant therapy with pegylated interferon alfa-2b versus observation alone in resected stage III melanoma: final results of EORTC 18991, a randomized phase III trial. *Lancet* **12:** 117–26.

18. Wolchok, J.D. *et al.* 2009. Guidelines for evaluation of immune therapy activity in solid tumors: immune-related response criteria. *Clin. Cancer Res.* **15:** 7412–7420.
19. Danielli, R. *et al.* Single-centre experience of thymosin alpha-1 and dacarbazine in advanced melanoma: results from an Italian compassionate-use programme. In preparation.
20. Di Giacomo, A.M. *et al.* 2011. Ipilimumab experience in heavely pretreated patients with melanoma in an expanded access program at the University Hospital of Siena (Italy). *Cancer Immunol. Immunother.* **60:** 467–477.
21. Romani, L. *et al.* 2006. Thymosin alpha1 act tryptophan catabolism and estabilishes a regulatory environment for balance of inflammation and tolerance. *Blood* **108:** 2265–2274.
22. Haynes, N.M. *et al.* 2008. Immunogenic anti-cancer chemotherapy as an emerging concept. *Curr. Opin. Immunol.* **20:** 545–557.
23. Krieg, A.M. *et al.* 2008. Toll-like receptor 9 (TLR9) agonists in the treatment of cancer. *Oncogene* **27:** 161–167.

Ann. N.Y. Acad. Sci. ISSN 0077-8923

ANNALS OF THE NEW YORK ACADEMY OF SCIENCES
Issue: *Thymosins in Health and Disease*

Thymosin α1 as a stimulatory agent of innate cell-mediated immune response

Annalucia Serafino,[1] Pasquale Pierimarchi,[1] Francesca Pica,[2] Federica Andreola,[1] Roberta Gaziano,[2] Noemi Moroni,[1] Manuela Zonfrillo,[1] Paola Sinibaldi-Vallebona,[2] and Enrico Garaci[2]

[1]Institute of Translational Pharmacology—National Research Council of Italy, Rome, Italy. [2]Department of Experimental Medicine and Surgery, University of Rome "Tor Vergata," Rome, Italy

Address for correspondence: Annalucia Serafino, Institute of Translational Pharmacology—National Research Council of Italy Via Fosso del Cavaliere 100,00133, Rome, Italy. annalucia.serafino@ift.cnr.it

The innate immune response and its cellular effectors—peripheral blood mononuclear cells and differentiated macrophages—play a crucial role in detection and elimination of pathogenic microorganisms. Chemotherapy and some immunosuppressive drugs used after organ transplantation and for treatment of autoimmune diseases have, as main side effect, bone marrow suppression, which can lead to a reduced response of the innate immune system. Hence, many immune-depressed patients have a higher risk of developing bacterial and invasive fungal infections compared with immune-competent individuals. Thymosin α1 (Tα1) immunomodulatory activity on effector cells of the innate immunity has been extensively described, even if its mechanism of action is not completely understood. Here, we report some of the main knowledge on this topic, focusing on our *in vitro* and *in vivo* work in progress that reinforce the validity of Tα1 as a stimulatory agent for detection and elimination of pathogens by differentiated macrophages and for restoring immune parameters after chemotherapy-induced myelosuppression.

Keywords: thymosin α1; macrophages; phagocytosis; myelosuppression

Introduction

Thymosin α1 (Tα1), in the form of a synthetic 28-amino-acid peptide,[1,2] is currently in clinical trials worldwide for the treatment of chronic B and C hepatitis[3,4] and cancer.[5] Furthermore, it has as additional indications for primary immune deficiencies,[6] HIV/AIDS,[7,8] and defective response to vaccinations.[9,10]

Tα1 immunomodulatory activity on cells of the innate immune system, including polymorphonucleates, dendritic cells, and macrophages has been extensively described, even if the mechanism/s of action of the peptide remain not yet completely understood. Aiming to deepen the knowledge on the causal mechanisms of Tα1 action on immune defense, current research in our laboratory has focused on the effects of this peptide on the activity of some cellular component of the innate immune system, both *in vitro* and *in vivo*. Particular attention has been given to Tα1 efficacy on the phagocytic and killing ability of human monocyte–derived macrophages (MDMs) against fungal pathogens and on the activity of peripheral blood mononuclear cells (PBMCs) after chemically induced bone marrow suppression in immunocompetent rats.

Immunomodulatory efficacy of Tα1: influence on phagocytosis in fungal infection

The monocytic/granulocytic system as well as differentiated macrophages constitute the primary cellular effectors of the immune response, playing a pivotal role in the detection and elimination of foreign bodies such as pathogenic microorganisms. Recognition of foreign microorganisms by these cells ultimately results in phagocytosis and in the eventual destruction of pathogens by lysosomal enzymes. Phagocytosis is an intrinsically complex process that involves various signalings whose synchronous activation

doi: 10.1111/j.1749-6632.2012.06707.x

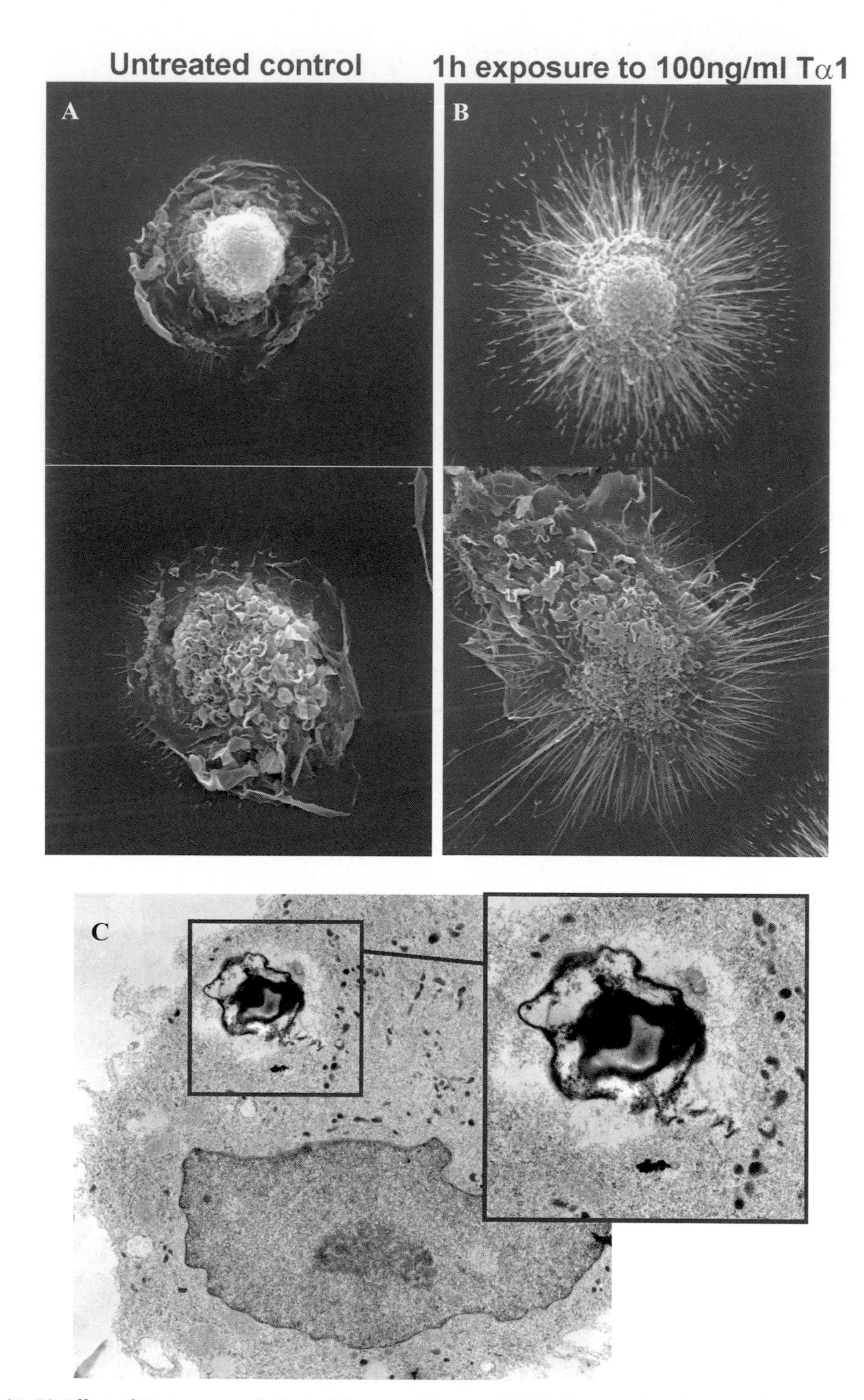

Figure 1. (A, B) Effect of Tα1 on morphological features of human MDMs by scanning electron microscopy (SEM). Human monocyte–derived macrophage cultures were obtained from peripheral blood mononuclear cells of healthy donors by density gradient centrifugation. Differentiated macrophages were isolated from the lymphocytic/monocytic fraction through adhesion to culture flasks. After one hour of *in vitro* exposure to 100 ng/mL Tα1 (B) MDMs exhibited morphological features typical of activated macrophages, showing more numerous microvillous structures and a rougher surface with more prominent filopodia, blebs, and ruffling compared to untreated controls (A). (C) The influence of Tα1 on the macrophage ability in killing of fungal conidia. Transmission electron microscopy images of 100 ng/mL Tα1-treated MDMs six hours postchallenge with *Aspergillus niger* conidia: detail of ingested conidia partially digested inside a lysosome. These data represent work in progress.

results in cytoskeleton rearrangement, alterations in membrane trafficking, particle engulfment, pathogen killing, and release of chemical mediators, that is, growth factors, pro- and anti-inflammatory cytokines and chemokines that guide the adaptive immune response.[11]

In the past, our group contributed in defining the therapeutic potential of Tα1, by several preclinical studies aimed at evaluating the immune modulating activity of this peptide when used in combination with cytokines and/or chemotherapeutic agents.[5,12–16] Tα1, together with other immune stimuli, has been reported to enhance natural killer cell activity and to induce the differentiation and maturation of T cells.[12,17] We also showed that Tα1 was able to protect untreated or cyclophosphamide-pretreated recipient mice from intravenous challenge with *Candida* albicans.[18] This effect was associated with a significant increase of the number of circulating polymorphonucleates (PMNs) as well as of their candidacidal activity, but the issue of whether Tα1 induced the stimulation of PMN precursor cells or if other cell populations, such as lymphocytes or monocytes, function as mediators remained unresolved.[19]

Other studies described the capacity of Tα1 to potentiate several murine as well as human macrophage skills.[20–22] Furthermore, Tα1 upregulates the expression of the major histocompatibility complex (MHC) class-I in primary cultures of human macrophages[23] and of Toll-like receptor (TLR) 2, 5, 8, and 9, also protecting mice from challenge by invasive aspergillosis in a myeloid differentiating factor 88 (MyD88)-dependent way.[24] Tα1 can also induce T cell and dendritic cell (DC) maturation as well as increase interleukin (IL)-12 expression.[24,25]

Even if these studies provide relevant insights into how Tα1 exerts its immunomodulatory

Table 1. ***In vitro*** **effect of Tα1 on phagocytic activity of human MDMs (work in progress)**

MDM phagocytic ability against nonspecific stimulus (polystyrene beads)*

Sample	% of phagocytic cells	No. of beads/cell	Significance** versus untreated control (t test, $P < 0.05$)
Untreated control	80.7	1.1 ± 0.28	
Tα1 50 ng/mL (one hour pretreatment)	79.58	13.2 ± 0.64	0.001586148
Tα1 100 ng/mL (one hour pretreatment)	94.45	29.5 ± 1.06	0.000718112

MDM phagocytic ability against *Apergillus niger conidia****

Sample	% of phagocytic cells	No. of conidia/cell	Significance** vs. untreated control (t test, $P < 0.05$)
Untreated control	4.1	0.15 ± 0.03	
Tα1 50 ng/mL (one hour pretreatment)	71.2	1.76 ± 0.12	0.0027
Tα1 100 ng/mL (one hour pretreatment)	84.5	3.7 ± 0.09	0.00036

*The phagocytic activity of treated and untreated MDMs was tested one hour after addition to the cultures yellow–green fluorescent polystyrene beads. Samples were observed by confocal microscopy and the number of phagocytic MDMs (reported as the percentage of phagocytic cells), as well as the number of beads per cell, was counted.
**Refer to the number of beads/cell.
***The engulfment of *A. niger* conidia by human MDMs has been assessed 2.5 hours postchallenge.

Table 2. *In vitro* effect of Tα1 on proteolytic and killing activity of human MDMs against *A. niger conidia* (work in progress)

Time	Sample	No. of *A. niger* conidia/infected cell	% of *A. niger* conidia in proteolytical compartments*
Six hours postchallenge	Untreated control	9.6	19.6
	Tα1 100 ng/mL	22.4	52.8
Removal of conidia from cell culture medium			
24 hours postchallenge (18 hours in conidia-free medium)	Untreated control	16.9	57.6
	Tα1 100 ng/mL	11.8	14.6

*The quantitative assessment was carried out by confocal microscopy after the staining of proteolytically active intracellular compartments with the specific fluorescent DQ-Red BSA. Results are reported as percentage of internalized conidia colocalizing with DQ-BSA positive compartments.

activity, additional research is needed for a complete comprehension of the mechanisms involved in the Tα1/cell interaction, particularly as it concerns the Tα1-mediated effects on pathogen engulfment and killing.

Preliminary results from ongoing studies performed in our lab show that after *in vitro* exposure to Tα1 human monocyte–derived macrophages (MDMs) undergo early morphological activation, including an enlarged size, numerous microvillous structures, and a rougher surface with more prominent filopodia, blebs, and ruffling, compared to untreated controls (Fig. 1A and B). Under these experimental conditions, Tα1 is able to increase, in a dose-dependent manner, the phagocytic activity of MDMs not only against a nonmicrobic foreign particles, such as fluorescent beads and the mean number of phagocytosed particles per cell, but also against fungal pathogens such as the *Aspergillus niger* conidia (Table 1), starting as soon as 30 min and up to six hours post-challenge.

As extensively demonstrated for dendritic cells,[24,26] our data indicate that Tα1 also influences the killing ability of MDMs against fungal conidia, suggested by the observation that at six hours postchallenge the majority of ingested conidia appear to be partially digested inside the lysosomes of Tα1-stimulated MDMs (Fig. 1C). This phenomenon seems to be associated with an augmented MDM ability of processing and disassembly the fungal pathogen, as indicated by the higher percentage of internalized conidia colocalizing with proteolytically active compartments in Tα1-stimulated cultures, compared with the untreated control (Table 2). The Tα1-stimulated killing ability ultimately results in a decrease in the number of internalized conidia after additional 18 hours of culture in conidia-free medium (Table 2). Based on these preliminary results, the Tα1-induced stimulation of the macrophage functions seems to be a highly efficient process occurring in a short time (hours).

Our current work in progress focuses on the early events associated with the Tα1-stimulated ability of macrophages to internalize, process, and disassemble fungal pathogens in order to clarify the mechanism of action of this immunostimulatory agent. We also discuss preliminary evidence indicating that Tα1 might implement pathogen internalization by means of a complex reorganization of selected cytoskeletal components, and with morphofunctional features that are reminiscent of the complement receptor-mediated phagocytosis (Serafino *et al.*, unpublished data).

Thymosin α1 efficacy in restoration of chemotherapic-induced myelotoxicity

Myelosuppression is a severe side effect of chemotherapy as well as of some immunosuppressive drugs used after organ transplantation and for treatment of autoimmune diseases. As a consequence of drug-induced bone marrow suppression, immune-depressed patients have a higher

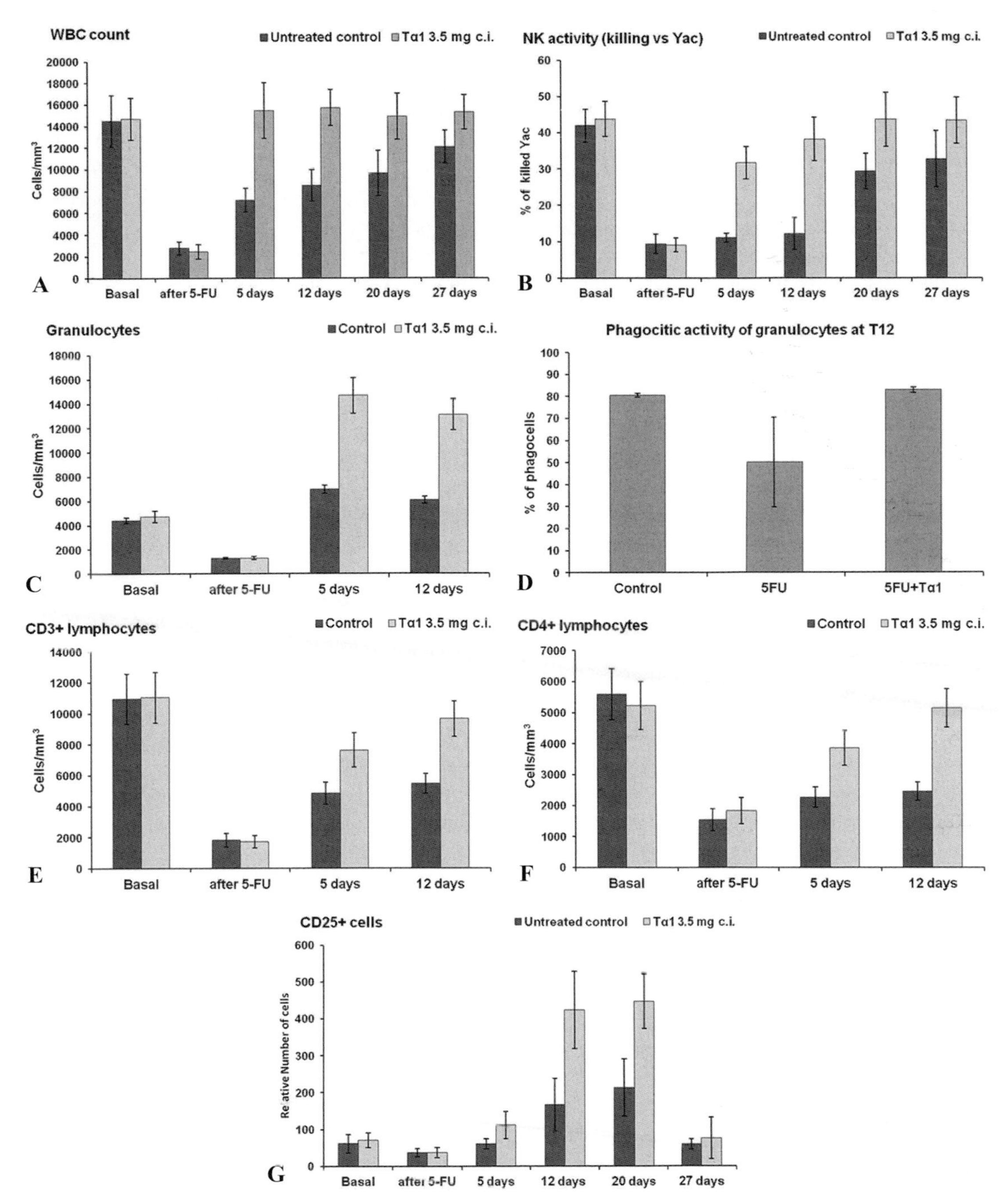

Figure 2. Results from the experiments on the effect of Tα1 administration to 5-FU myelosuppressed BDIX rats. Animals, 10 weeks old and weighing 220–250 g, received 100 mg/kg 5-fluorouracil (5-FU) for immune suppression. Eight days after 5-FU treatment, rats were randomly assigned to one of the following groups ($n = 8$): (1) control (saline in osmotic mini-pumps); (2) continuous infusion (c.i.) of Tα1, administered by osmotic mini-pumps at 3.5 mg/kg/5 days. Tα1 continuous infusion was obtained using an osmotic mini-pump implanted subcutaneously. All animals were implanted with pumps filled with Tα1 or saline as appropriate. The evaluation of the immune parameters included total leukocyte (WBC) and granulocyte number (judged by flow cytometry, based on forward and sideward scatter patterns); granulocyte phagocytic ability (by flow cytometry using a phagotest kit); activated lymphocytes and monocytes ($CD3^+$,$CD4^+$, and $CD25^+$ cells by flow cytometry); and NK activity (LDH released from YAC-1 cells after 4-h exposure to PBMC). These data represent work in progress.

risk of developing bacterial and invasive fungal infections compared with immune-competent individuals. The frequency of invasive mycoses due to opportunistic fungal pathogens has increased significantly over the past decades.

The protective activity of Tα1 against opportunistic infections in animal models was first described in the early 1980s.[27] More recently, a phase I/II clinical study demonstrated that Tα1 administration is able to significantly improve the polymorphonuclear (phagocytosis) and dendritic cell (phagocytosis, expression of costimulatory molecules, and cytokine production) functions in recipients of haploidentical stem cell transplants for hematologic malignancies.[28] This study also reported an association of Tα1 administration with increased T cell counts and earlier appearance of functional pathogen-specific T cell responses (against *Aspergillus*, *Candida*, cytomegalovirus, varicella–zoster virus, herpes simplex virus, and *Toxoplasma*).

In preliminary experiments performed in our lab, using a rat model of chemotherapy-induced myelosuppression, we obtained evidence that Tα1 treatment is able to induce a rescue of the number and activity of peripheral blood mononuclear cell (PBMC) at near-normal values. The animal model developed and used by our group consists of syngeneic immunocompetent rats (inbred male BDIX rat strain) in which bone marrow suppression is induced by intraperitoneal injection of a single dose of 100 mg/kg of 5-fluorouracil (5-FU). This is a chemotherapeutic drug extensively used for treatment of different types of cancer,[29,30] including bowel, breast, skin, stomach, and esophageal cancers, that produce myelotoxicity as side effect.[31] In our preliminary experiments, Tα1 was administered by continuous infusion implanted subcutaneously. The modality of treatment via continuous infusion was adopted on the basis of our previous results demonstrating that maintenance of a constant level of Tα1 over five days in the circulation increases immunological effects, compared to the subcutaneous injection of a single dose (Pierimarchi *et al.*, unpublished data). The immune parameters were determined with PBMC from blood collected by intracardiac puncture at baseline (day 0–T_0) and eight days after 5-FU treatment (day 1 of Tα1 treatment), and monitored during Tα1 administration at 5, 12, 20, and 27 days treatment. Results obtained and summarized in Figure 2, demonstrated that Tα1 administration by continuous infusion was able to restore the total leukocyte number as well as NK activity and the number of $CD3^+$ and $CD4^+$ lymphocytes to values near to that of untreated immunocompetent rats (Fig. 2A, B, E, and F), in accordance with what has previously reported for other *in vitro* and *in vivo* experimental models.[12,17] Moreover, in immune-suppressed rats, Tα1 also augmented the percentage of circulating granulocytes and induced a recovery of granulocyte phagocytic ability, that are two parameters significantly reduced by 5-FU treatment (Fig. 2C and D). We also examined the PBMC fraction for the expression of surface molecules that may be indicative of leucocytes activation. In particular, we are using flow cytometry to evaluate the expression of CD25, a marker of circulating monocyte/lymphocyte activation.[32,33] Tα1 treatment induced a significant increase of the percentage of $CD25^+$ cells in circulating cells compared to the untreated control rats, and this effect is particularly evident at 12–20 days of Tα1 administration (Fig. 2G).

Conclusions and remarks

Data have been published demonstrating that Tα1 is an immunoregulatory agent promoting the coordinated activation of the innate and adaptive T helper immunity. Our *in vitro* and *in vivo* findings detailed in this review reinforce the validity of this thymic peptide as a stimulatory agent of innate cell-mediated immune response, particularly as it concerns its abilities to implement detection and elimination of fungal pathogens by differentiated macrophages and to restore to near-normal values the hematological and immune parameters depleted after chemotherapy-induced bone marrow suppression. This supports the therapeutic utility of Tα1 as a promising adjuvant candidate in immunosuppressive pathologies and in infectious diseases. Further, the *in vivo* findings indicate that continuous infusion might have a role in the efficacy of Tα1 treatment, and that this administration regimen should be taken into account in the clinical setting.

Acknowledgments

This work was supported by the following grants from the Italian Ministry of University and Research (MIUR): Research Projects of National Interest

(PRIN) 2008 to EG; MIUR Prot. 10484 to PSV and EG; MIUR Prot. 1558 to PP. We would like to thank Pamela Papa and Martino T. Miele for administrative support, and Matilde Paggiolu and Giuseppe Nicotera for scientific secretarial assistance.

Conflicts of interest

The authors declare no conflicts of interest.

References

1. Low, T.L. & A.L. Goldstein. 1979. The chemistry and biology of thymosin. II. Amino acid sequence analysis of thymosin alpha1 and polypeptide beta1. *J. Biol. Chem.* **254:** 987–995.
2. Wang, S.S., R. Makofske, A. Bach & R.B. Merrifield. 1980. Automated solid phase synthesis of thymosin alpha 1. *Int. J. Pept. Protein Res.* **15:** 1–4.
3. Rasi, G., D. Di Virgilio, M.G. Mutchnick, *et al.* 1996. Combination thymosin alpha 1 and lymphoblastoid interferon treatment in chronic hepatitis C. *Gut* **39:** 679–683.
4. Rasi, G., M.G. Mutchnick, D. Di Virgilio, *et al.* 1996. Combination low-dose lymphoblastoid interferon and thymosin alpha 1 therapy in the treatment of chronic hepatitis B. *J. Viral Hepat.* **3:** 191–196.
5. Garaci, E., C. Favalli, F. Pica, *et al.* 2007. Thymosin alpha 1: from bench to bedside. *Ann. N.Y. Acad. Sci.* **1112:** 225–234.
6. Wara, D.W., A.L. Goldstein, N.E. Doyle & A.J. Amman. 1975. Thymosin activity in patients with cellular immunodeficiency. *N. Engl. J. Med.* **292:** 70–74.
7. Billich, A. 2002. Thymosin alpha1, SciClone Pharmaceuticals. *Curr. Opin. Investig. Drugs.* **3:** 698–707.
8. Garaci, E., G. Rocchi, L. Perroni, *et al.* 1994. Combination treatment with zidovudine, thymosin alpha 1 and interferon-alpha in human immunodeficiency virus infection. *Int. J. Clin. Lab. Res.* **24:** 23–28.
9. Ershler, W.B., S. Gravenstein & Z.S. Geloo. 2007. Thymosin alpha 1 as an adjunct to influenza vaccination in the elderly: rationale and trial summaries. *Ann. N.Y. Acad. Sci.* **1112:** 375–384.
10. Goldstein, A.L. & M. Badamchian. 2004. Thymosins: chemistry and biological properties in health and disease. *Expert. Opin. Biol. Ther.* **4:** 559–573.
11. Underhill, D.M. & A. Ozinsky. 2002. Phagocytosis of microbes: complexity in action. *Annu. Rev. Immunol.* **20:** 825–852.
12. Favalli, C., T. Jezzi, A. Mastino, *et al.* 1985. Modulation of natural killer activity by thymosin alpha 1 and interferon. *Cancer Immunol. Immunother.* **20:** 189–192.
13. Garaci, E., A. Mastino, F. Pica & C. Favalli. 1990. Combination treatment using thymosin alpha 1 and interferon after cyclophosphamide is able to cure Lewis lung carcinoma in mice. *Cancer Immunol. Immunother.* **32:** 154–160.
14. Garaci, E., G. Rocchi, L. Perroni, *et al.* 1994. Combination treatment with zidovudine, thymosin alpha 1 and interferon-alpha in human immunodeficiency virus infection. *Int. J. Clin. Lab. Res.* **24:** 23–28.
15. Garaci, E., F. Pica, G. Rasi & C. Favalli. 2000. Thymosin alpha1 in the treatment of cancer: from basic research to clinical application. *Int. J. Immunopharmacol.* **22:** 1067–1076.
16. Mastino, A., C. Favalli, S. Grelli, *et al.* 1992. Combination therapy with thymosin alpha 1 potentiates the anti-tumor activity of interleukin-2 with cyclophosphamide in the treatment of the Lewis lung carcinoma in mice. *Int. J. Cancer.* **50:** 493–499.
17. Sztein, M.B., S.A. Serrate & A.L. Goldstein. 1986. Modulation of interleukin 2 receptor expression on normal human lymphocytes by thymic hormones. *Proc. Natl. Acad. Sci. USA* **83:** 6107–6111.
18. Bistoni, F., P. Marconi, L. Frati, *et al.* 1982. Increase of mouse resistance to Candida albicans infection by thymosin alpha 1. *Infect. Immun.* **36:** 609–614.
19. Bistoni, F., M. Baccarini, E. Blasi, *et al.* 1985. Modulation of polymorphonucleate-mediated cytotoxicity against Candida albicans by thymosin alpha 1. *Thymus* **7:** 69–84.
20. Hu, S.K., M. Badamchian, Y.L. Mitcho & A.L. Goldstein. 1989. Thymosin enhances the production of IL-1 alpha by human peripheral blood monocytes. *Lymphokine Res. Fall.* **8:** 203–214.
21. Shrivastava, P., S.M. Singh & N. Singh. 2004. Activation of tumor-associated macrophages by thymosin alpha 1. *Int. J. Immunopathol. Pharmacol.* **17:** 39–47.
22. Tzehoval, E., M.B. Sztein & A.L. Goldstein. 1989. Thymosins apha 1 and beta 4 potentiate the antigen-presenting capacity of macrophages. *Immunopharmacology* **18:** 107–113.
23. Giuliani, C., G. Napolitano, A. Mastino, *et al.* 2000. Thymosin-alpha1 regulates MHC class I expression in FRTL-5 cells at transcriptional level. *Eur. J. Immunol.* **30:** 778–786.
24. Romani, L., F. Bistoni, R. Gaziano, *et al.* 2004. Thymosin alpha 1 activates dendritic cells for antifungal Th1 resistance through toll-like receptor signaling. *Blood* **103:** 4232–4239.
25. Knutsen, A.P., J.J. Freeman, K.R. Mueller, *et al.* 1999. Thymosin-alpha1 stimulates maturation of CD34+ stem cells into CD3 + 4+ cells in an in vitro thymic epithelia organ coculture model. *Int. J. Immunopharmacol.* **21:** 15–26.
26. Romani, L., F. Bistoni, C. Montagnoli, *et al.* 2007. Thymosin alpha1: an endogenous regulator of inflammation, immunity, and tolerance. *Ann. N.Y. Acad. Sci.* **1112:** 326–338.
27. Ishitsuka, H., Y. Umeda, J. Nakamura & Y. Yagi. 1983. Protective activity of thymosin against opportunistic infections in animal models. *Cancer Immunol. Immunother.* **14:** 145–150.
28. Perruccio, K., P. Bonifazi, F. Topini, *et al.* 2010. Thymosin alpha1 to harness immunity to pathogens after haploidentical hematopoietic transplantation. *Ann. N.Y. Acad. Sci.* **1194:** 153–161.
29. Heidelberg, C. & F.J. Ansfield. 1963. Experimental and clinical use of fluorinated pyrimidines in cancer chemotherapy. *Cancer Res.* **23:** 1226–1243.
30. Seifert, P., H.L. Baker & M.L. Reed. 1975. Comparison of continuously infused 5-fluorouracil with bolus injection in treatment of patients with colorectal adenocarcinoma. *Cancer* **36:** 123–128.
31. Schetz, J.D., H.J. Wallance & R.B. Diasio. 1984. 5-Fluorouracil incorporation into DNA of CF-1 mouse bone marrow cells as a possible mechanism of toxicity. *Cancer Res.* **44:** 1358–1363.

32. Espinoza-Delgado, I., D.L. Longo, G.L. Gusella & L. Varesio. 1992. Regulation of IL-2 receptor subunit genes in human monocytes. Differential effects of IL-2 and IFN-gamma. *J. Immunol.* **149:** 2961–2968.
33. Sanarico, N., A. Ciaramella, A. Sacchi, *et al.* 2006. Human monocyte-derived dendritic cells differentiated in the presence of IL-2 produce proinflammatory cytokines and prime Th1 immune response. *J. Leukocyte Biol.* **80:** 555–562.

Ann. N.Y. Acad. Sci. ISSN 0077-8923

ANNALS OF THE NEW YORK ACADEMY OF SCIENCES
Issue: *Thymosins in Health and Disease*

Thymosin α1 continues to show promise as an enhancer for vaccine response

Cynthia Tuthill,[1] Israel Rios,[1] Alfonso De Rosa,[2] and Roberto Camerini[2]
[1]SciClone Pharmaceuticals, Inc., Foster City, California. [2]Sigma-Tau S.p.A., Pomezia, Italy

Address for correspondence: Cynthia Tuthill, Senior Vice President, Scientific Affairs and Chief Scientific Officer, SciClone Pharmaceuticals, Inc., 950 Tower Lane, Foster City, CA 94404. ctuthill@sciclone.com

Thymosin α1 (Tα1) is an immune-modulating peptide that can be expected to improve response to vaccinations, as stimulated dendritic cells and T cells can act in concert to increase antibody production along with an improved cytotoxic response from the T cells themselves. Tα1 demonstrated efficacy in preclinical studies; subsequently, it was shown to enhance response to vaccinations in difficult-to-treat populations, including individuals immune suppressed due to age or hemodialysis, and leading to a decrease in later infections. During the 2009 pandemic outbreak of H1N1 influenza, mouse and ferret studies confirmed that the use of higher doses of Tα1 allowed for fewer injections than those used in the previous clinical studies. In addition, a clinical study with Focetria™ MF59-adjuvanted monovalent H1N1 vaccine showed that treatment with Tα1 twice provided an earlier and greater response to the vaccine ($P < 0.01$).

Keywords: thymosin; thymalfasin; Zadaxin; vaccine; influenza

Introduction

As exemplified by the 2009 pandemic outbreak of H1N1 influenza and recent concerns over the possible spread of H5N1 influenza, there is an urgent need for novel vaccine-enhancing compounds to expand an available vaccine supply through dose sparing. Such a vaccine enhancer could also broaden the response to the vaccine across multiple virus strains. A further opportunity exists due to the significant numbers of individuals who historically have been shown to be refractory to vaccines, including those immune suppressed by underlying disease or being treated with chemotherapy or invasive procedures such as hemodialysis, and the elderly.

Thymosin α1 (Tα1; trade name Zadaxin®) is a peptide that has been evaluated for therapeutic potential in several conditions and diseases, not only for treatment of cancer and chronic viral infection, but also for depressed response to vaccination.

Mechanism of action of Tα1 for vaccine enhancement

Investigation of the mechanism of action of Tα1 at the cellular level has implicated a number of intracellular cell-signaling pathways leading to stimulation of the immune system. These immunological effects can explain the effectiveness of Tα1 in enhancement of vaccines, as well as other indications where a stimulated or enhanced immune response is important for an improved host response, including cancer and viral infections.

Tα1 has been shown to be a toll-like receptor (TLR)9 agonist.[1,2] The TLRs are a family of proteins that mediate innate immunity; stimulation of one or more TLRs by a TLR agonist can enhance the adaptive immune response, which is critical for humoral immunity. Tα1 affects both myeloid and plasmacytoid dendritic cells (DCs), the professional antigen-presenting cells, leading to activation and stimulation of signaling pathways that initiate production of immune-related cytokines.[2] Tα1 also affects precursor T cells, leading to an increase in the number of activated T helper (Th) cells ($CD4^+$ T cells)[3,4] and shifting toward the Th1 subclass.[5,6] This shift leads to increased expression of Th1-type cytokines such as interleukin (IL)-2 (Refs. 7–10) and IFNα.[7–13] The activated DCs and Th1 cells can then act in concert, leading to stimulation of differentiation of specific B cells to antibody-producing plasma cells and an improvement in response to vaccines by stimulation of antibody production[14–17] (Fig. 1A).

doi: 10.1111/j.1749-6632.2012.06680.x

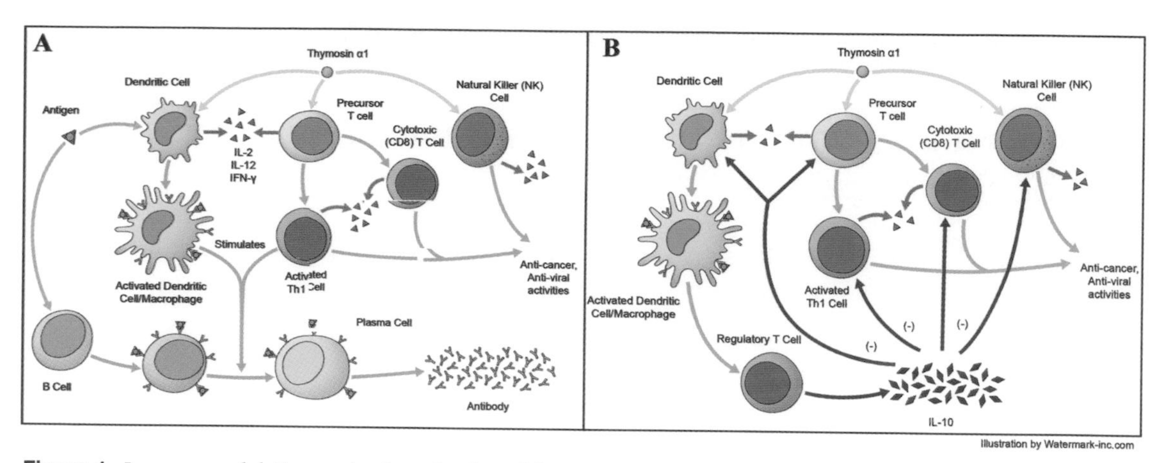

Figure 1. Immune-modulating mechanism of action of thymosin α1. (A) Tα1 interacts with TLR9 in dendritic and T cells, leading to stimulation of cellular subsets, which act in concert to increase cytotoxic T cell activity and production of antibodies in response to vaccines. (B) Tα1 also stimulates the activity of regulatory T cells, leading to a dampening of the immune response and preventing a cytokine storm or overinflammation.

The effects of Tα1 on TLR9 lead to stimulation of the NF-κB and p38 MAPK pathways,[4,18,19] both of which play critical roles in the maturation of DCs[20,21] and in antigen presentation by DCs.[22] Tα1 treatment leads to increased expression of the thymopoetic cytokines IFNα, IL-7, and IL-15.[1,23]

Importantly, it has also been shown that Tα1 stimulates activity of indoleamine-2,3-dioxygenase (IDO) in plasmacytoid DCs.[1,24,25] Stimulation of IDO leads to an increase in FoxP3$^+$ IL-10–producing regulatory T cells, and this increase leads to feedback inhibition of cytokine production, hence dampening immune response to prevent a proinflammatory cytokine storm and possibly autoimmune phenomena (Fig. 1B).

Safety of Tα1

This mechanism of action could be the basis for the excellent safety record for SciClone's Zadaxin® Tα1, which has been approved in over 35 countries, is well tolerated, and not been associated with any significant side effects in the postmarketing treatment experience of more than 300,000 patients. Tα1 has been administered without adverse incidents to elderly subjects (up to 101 years old), children (as young as 13 months), and immunocompromised patients. Thus, while Tα1 is one of only a few immunomodulators that have been approved for human use, it does not appear to induce any of the side effects and toxicities commonly associated with agents in this class such as interferon and IL-2.

Several clinical studies evaluated Tα1 as a monotherapy treatment and reported levels of circulating immune cells. Tα1 treatment led to small increases in lymphocytes, natural killer (NK) cells, IFN-γ synthesis, and overall T cell levels during treatment for lung cancer,[26] hepatocellular carcinoma,[27] hepatitis B,[13] and for vaccination with the hepatitis B vaccine.[28] However, none of these studies led to an irreversible or overtly inflammatory immune system response, supporting the safety profile of this compound.

Tα1 as a vaccine enhancer

As part of the normal aging process, a gradual decline in thymus function and thymic hormone production is seen. The decrease in available circulating thymic hormones may contribute to the decline in immune function, particularly the T cell component.[29–31] In the elderly, antibody responses after vaccination have been shown to be impaired when compared with response in young subjects.[15,32]

This decreased antibody response to T cell–dependent antigens, particularly in the elderly, may be one factor that accounts for insufficient efficacy of certain vaccination programs (e.g., influenza). Diminished antibody responses have also been reported in patients with end-stage renal disease. The evidence for impairment of cell-mediated immunity in hemodialysis patients has been attributed to incompetence in T cell–mediated immune responses.[29,33–37] Several studies have reported poor antibody response after hepatitis B vaccination in hemodialysis patients.[38–40]

Because Tα1 can enhance T cell–dependent specific antibody production, the addition of Tα1 to vaccination protocols for immunocompromised individuals should be effective. Evaluation of Tα1 for this use was first investigated in several preclinical studies. The response to tetanus toxoid (TT) vaccine was evaluated in old (23 months) versus young (two to three months) mice; the old mice had a significantly lower antibody response than young mice.[14] The effect of treatment with Tα1, at a dose of 0.5 ug/kg given on the day of vaccination and daily for four additional days after vaccination, was to improve the response; production of anti-TT antibody was increased significantly ($P < 0.05$) in both young and old mice. In fact, the treatment with Tα1 restored the antibody response in the old mice to levels seen in the young animals.

A similar effect of Tα1, to improve the response of older mice to the levels seen in younger mice, was also obtained with influenza vaccine. In this study, mice were pretreated with a much higher dose of Tα1 (10 μg/day or approximately 0.5 mg/kg for five days), and virus-specific cytotoxic response was determined.[15] Tα1 treatment increased the vaccine response of the old (24–26 months) mice to the levels seen in the young (2–6 months) mice.

Enhancement of specific cytotoxic antibody responses was also observed by the administration of Tα1 to mice immunosuppressed with cocaine.[16] In this study, mice were given cocaine injections for five days; saline or Tα1 at a dose of 0.2 mg/kg were given each day during cocaine administration and for an additional three days. Spleen cells were evaluated for specific cytotoxic activity one day after completion of treatment and Tα1 led to complete protection against the induced lack of T cell responsiveness.

In vitro antibody synthesis has also been shown to be augmented by Tα1. Peripheral blood lymphocytes (PBL) isolated from human subjects who had been vaccinated against seasonal influenza were incubated with or without Tα1.[17] As expected, in this study elderly individuals (over 65 years) showed a lower *in vivo* antibody response than younger persons (less than 30 years). PBLs were isolated from these immunized young and old subjects and incubated with various thymic peptides *in vitro*. Tα1 at doses of 0.01, 0.1, or 1.0 ng/mL amplified the antibody response in PBL from 16 of the 28 elderly subjects (57%), compared to seven of the 30 young subjects (23%). Treatment with the highest dose of 1 ng/mL Tα1 resulted in twice as many responders as the lowest dose (6 vs. 3), suggesting a possible dose response. This is a particularly intriguing finding, as circulating levels of endogenous Tα1 in healthy individuals is about 1 ng/mL.

These experimental data support that Tα1, when appropriately administered with a vaccine, can enhance immune function and enhance an antibody response in individuals with a compromised immune system.

Building on these promising preclinical results, Tα1 was evaluated in several large clinical trials in the 1980s for its ability to enhance response to vaccinations in historically difficult-to-treat populations, including persons who were immune-suppressed due to age or hemodialysis (Table 1).[26,41–45,47] As described in detail later, these studies demonstrated the effectiveness of Tα1 in increasing specific antibody response to both the seasonal influenza and hepatitis B vaccinations, and a decrease in

Table 1. Summary of clinical studies with thymosin α1 in vaccine enhancement

Study population	*n*	Vaccine	Results: Vaccine + Tα1	Results: Vaccine alone
Elderly subjects (pilot study)[41]	9	Influenza	67%	10% (historical)
Elderly male veterans[42]	90	Influenza	69%	52%
Elderly male veterans[43]	330	Influenza	70% (5% influenza cases)	35% (19% influenza cases)
Hemodialysis patients[44]	97	Influenza	65%	24%
Hemodialysis patients[26,45]	23	Hepatitis B	64%	17%
Hemodialysis patients[47]	92	Adjuvanted influenza	88%/89%	56%

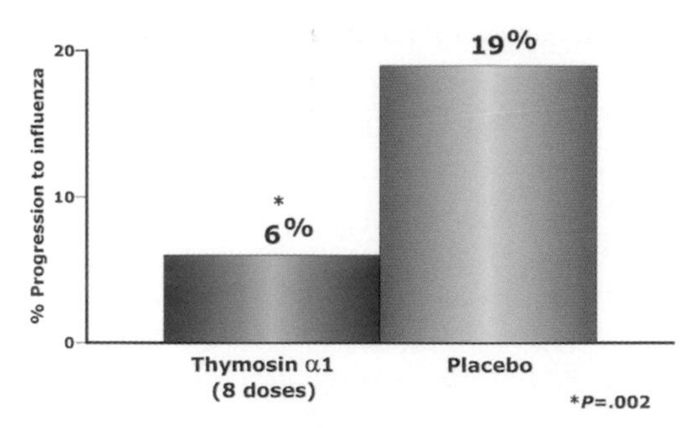

Figure 2. Lower incidence of influenza in subjects who received thymosin α1 after vaccination. Subjects treated with Tα1 at a dose of 0.9 mg/m^2 twice weekly for four weeks had a significantly decreased incidence of influenza later in the season, from 19% in patients receiving influenza vaccine with placebo to 6% in patients receiving vaccine in combination with Tα1 ($P = 0.002$).

subsequent influenza infections was even seen in one of the studies (Fig. 2).[43]

The immune-enhancing effect of Tα1 for influenza vaccination was first examined in a pilot study by Gravenstein *et al.* at the University of Wisconsin and Cornell Medical Center.[41] In this pilot trial, the effect of Tα1 on influenza vaccination was evaluated in nine elderly subjects (age range: 65–99 years) who had been nonresponsive to influenza vaccination the previous year. Tα1, at a dose of 0.9 mg/m^2 (approximately 1.6 mg for a person of 70 kg) was given twice per week for five weeks following a single injection of seasonal influenza vaccine. High levels of anti-influenza antibodies were seen in 67% (6/9) of the subjects, compared to a historical rate of 10% after revaccination in elderly individuals.

This pilot trial was followed by a double-blind, randomized, placebo-controlled study conducted by the same researchers.[42] Ninety male veterans at the Wisconsin Veterans' Administration Medical Center in Madison, Wisconsin, who were over 64 years of age (range 65–99 years) were randomized to receive either Tα1 (45 subjects) at a dose of 0.9 mg/m^2 or placebo (45 subjects) twice a week for four weeks following injection with the seasonal trivalent influenza vaccine. Groups were similar with respect to age, underlying disease, and concomitant medications. Response was defined as a fourfold or greater rise in antibody titer over three–six weeks, as measured by ELISA. 69% (31/45) of Tα1-treated subjects were effectively immunized, compared to 52% (21/40) treated with placebo ($P = 0.023$). The differences seen were greater in subjects older than 77 years; the relationship between antibody levels and age was also significant ($P < 0.039$), with antibody levels decreasing with age in placebo-treated subjects but remaining stable in the Tα1 group. Similar to results obtained in the mouse studies, the antibody levels seen after treatment with Tα1 were comparable to those seen in younger subjects.

The effect of Tα1 on response to influenza vaccine was evaluated in an even larger study at George Washington University, where 330 elderly subjects were vaccinated with the trivalent seasonal influenza vaccine (B/Ann Arbor, A/H3N2 Leningrad, A/H1N1 Taiwan).[43] Tα1 was given twice a week at a dose of 0.9 mg/m^2 for either two weeks (120 subjects) or four weeks (100 subjects) after vaccination; placebo was given for two weeks (110 subjects). In the subjects treated with Tα1 for four weeks, greater antibody levels for H1N1 Taiwan were seen compared to placebo or those treated with only two weeks of Tα1 ($P = 0.015$). In addition to the improved antibody response, the subjects treated with Tα1 for four weeks also had a significantly decreased incidence of influenza, from 19% in patients receiving influenza vaccine with placebo to 6% in patients receiving vaccine in combination with Tα1 ($P = 0.002$; Fig. 2). In subjects over 80 years of age, the responses to all three strains of influenza were significantly greater in subjects treated with Tα1 for eight weeks than subjects treated with placebo ($P < 0.05$).

The influence of Tα1 on a different population of immunocompromised subjects, those undergoing hemodialysis for chronic renal failure, was evaluated in a study at the University of Maryland, Baltimore with 97 subjects given the monovalent A/Taiwan/1/86 (H1N1) vaccine.[44] Vaccination was followed by Tα1 at a dose of 0.9 mg/m^2 twice weekly for four weeks. Response was defined as a fourfold or higher titer of specific anti-influenza antibody measured four weeks after vaccination, and was seen in 71% (34/48) of subjects treated with Tα1 compared to 43% (21/49) for those treated with placebo after vaccination ($P < 0.002$). This effect was long lasting, as well; when evaluated eight weeks after vaccination the Tα1-treated subjects still had a response rate of 65%, compared to only 24% in the placebo-treated subjects ($P < 0.001$).

Finally, the effect of Tα1 on vaccination response was also evaluated in another group of subjects undergoing hemodialysis for chronic renal failure. In this placebo-controlled double-blind study conducted at the Division of Nephrology, University of Maryland, 23 previous nonresponders to a course of Heptavax® (Merck Sharp and Dohme Corp., Whitehouse Station, NJ, USA) hepatitis B vaccine were retreated with three Heptavax injections one month apart.[27,45] Tα1 at a dose of 0.9 mg/m^2 or placebo was given twice a week for five treatments after each vaccination. When response, defined as anti-hepatitis B surface antigen antibody (HBsAb) titers greater than 10 million international units, was measured three months later, 64% (7/11) of Tα1-treated subjects had clinically significant titers, compared to 17% (2/12) of placebo-treated subjects ($P < 0.002$). This highly significant effect of Tα1 was also long-lasting: measured after 12 months, 45% of the Tα1-treated subjects still had clinically significant titers, while none of the placebo-treated subjects retained theirs ($P < 0.002$). Evaluation of peripheral T cell subsets also showed an improvement after treatment with Tα1, in keeping with the mechanism of action of this immunomodulating peptide.

Table 2. **Immune response at various times measured by a hemagglutination-inhibition (HI) assay**

Groups	Vaccine	Vaccine + Tα1 (3.2 mg)	Vaccine + Tα1 (6.4 mg)
Number of subjects	32	28	32
Day 21 HI test			
Percent with SC[a]	53	89	88
Percent with HI ≥ 1:40	76	93	94
GMR[b]	3.8	11.7	16.8
Day 42 HI test			
Percent with SC[a]	71	93	94
Percent with HI ≥ 1:40	94	100	100
GMR[b]	7.5	12.6	15.3

NOTE: Data from Ref. 47.
[a]SC = seroconversion, defined as negative prevaccination serum (i.e., HI titer < 1:10) and postvaccination HI titer ≥ 1:40, or a fourfold increase from nonnegative (≥1:10) prevaccination HI titer.
[b]GMR = geometric mean ratio = ratios of day *x*/day 1 geometric mean HI titers.

Recent preclinical and clinical studies of Tα1 as vaccine enhancer

The treatment schedule for Tα1 in the clinical studies described earlier was based on the regimen in use at the time, with Tα1 at a dose of 0.9 mg/m^2 given twice a week for up to several months after vaccination. In quickly responding to the perceived threat from the virulent new strain of influenza in 2009, SciClone and Sigma-tau quickly undertook mouse, ferret, and clinical studies to determine effectiveness against this strain and to ascertain whether the use of higher doses of TA1 could allow for fewer injections than those previously evaluated. Before the availability of the 2009 vaccine itself, response to the previous year's seasonal influenza vaccine was tested in animal models to determine the optimal dosing schedule for the planned subsequent clinical study. Mice were treated with Fluvirin® (Novartis Vaccines and Diagnostics S.r.L., Siena, Italy) vaccine and ferrets with either MF59-adjuvanted (Fluad®; Novartis Vaccines and Diagnostics) or nonadjuvanted (Agrippal®; Chiron S.p.A., Siena, Italy) vaccine, and Tα1 was administered at a range of doses in various treatment regimens. Results from these studies showed that antibody titers were greater in mice and ferrets receiving Tα1; administration of 1.2 mg/kg to mice or 0.6 mg/kg to ferrets, both seven days before vaccination and again on the day of vaccination, was found to be the optimal dosing regimen, leading to improvement in response to all three strains of influenza.[46] The titers of antibody were greater when determined 21 days after vaccination, and persisted when evaluated 42 days after vaccination. The improvement in titers was seen even in animals given a vaccine booster on day 21.

Using these results for optimal dosing and regimen, a randomized, open-label, three parallel arm clinical study in subjects on chronic dialysis due to end-stage renal disease was conducted in the departments of nephrology and dialysis at Padova Hospital in Padoa, Italy.[47] Findings showed that, when measured 21 days following vaccination, 89% (25/28) of patients treated twice (seven days before vaccination and on the day of vaccination) with a 3.2-mg dose of Tα1, and 88% (28/32) of patients treated twice with a 6.4-mg dose seroconverted (a fourfold or greater change in titers, as measured by hemagglutination inhibition [HI titers], from baseline) compared to only 56% (18/32) in patients treated with the Focetria™ (Novartis Vaccines and Diagnostics) MF59-adjuvanted monovalent H1N1 vaccine alone. The response to the higher dose of Tα1 was a highly statistically significant increase ($P < 0.01$); not only did Tα1 treatment allow for a greater response, but this response was seen sooner and maintained at day 42 after vaccination. The same findings were observed for the rate of seroprotection (HI titer > 1:40) and geometric mean ratio (Table 2). Tolerability was excellent in all three treatment groups.

Conclusions

The immune-stimulating effects of Tα1, especially the fact that treatment leads to stimulation of both cellular and humoral responses, explain how this molecule can lead to an effective increase in antibody production after vaccination. This enhancement would be especially vital for special populations who have impaired response to vaccines, such as the elderly, those infected with HIV, organ-transplant recipients, or those on hemodialysis, as well as for use in the general population as a vaccine-sparing agent. Importantly, the use of Tα1 could result in a faster response to vaccination for a postevent prophylaxis. Zadaxin® (Patheon Italia, Monza, Italy) (Tα1) is one of only a few immunomodulators that has been approved for human use. Importantly, due to its unique mechanism of action, it has an excellent safety profile and does not appear to induce the side effects and toxicities commonly associated with agents in this class. Although the previous studies focused on influenza and hepatitis B vaccines, the mechanism of action of Tα1 positions it for future use in combination with any vaccine—for example, those for anthrax, which currently generate suboptimal responses even after multiple doses in the general population.

Conflicts of interest

The authors declare no conflicts of interest. C. Tuthill and I. Rios are employees of SciClone Pharmaceuticals, Inc.; A. De Rosa and R. Camerini are employees of Sigma-tau.

References

1. Romani, L., F. Bistoni, R. Gaziano, *et al.* 2004. Thymosin alpha 1 activates dendritic cells for antifungal Th1 resistance through toll-like receptor signaling. *Blood* **103:** 4232–4239.
2. Romani L., F. Bistoni, K. Perruccio, *et al.* 2006. Thymosin alpha1 activates dendritic cell tryptophan catabolism and establishes a regulatory environment for balance of inflammation and tolerance. *Blood* **108:** 2265–2274.
3. Peng, Y., Z. Chen, W. Yu, *et al.* 2008. Effects of thymic polypeptides on the thymopoiesis of mouse embryonic stem cells. *Cell Biol. Int.* **32:** 1265–1271.
4. Yao, Q., L.X. Doan, R. Zhang, *et al.* 2007. Thymosin-a1 modulated dendritic cell differentiation and functional maturation from peripheral blood CD14+ monocytes. *Immunol. Lett.* **110:** 110–120.
5. Gramenzi, A., C. Cursaro, M. Margotti, *et al.* 2008. *In vitro* effect of thymosin alpha 1 and interferon alpha on Th1 and Th2 cytokine synthesis in patients with HBEAg-negative chronic hepatitis B. *J. Viral Hepatitis* **15:** 442–448.
6. Cursaro, C., M. Margotti, L. Favarelli, *et al.* 1998. Thymosinα1 (Tα1) plus interferonα (IFNα) enhance the immune and antiviral response of patients with hepatitis C virus infection. *Hepatology* **28:** 361A.
7. Serrate, S., R. Schulof, L. Leondaridis, *et al.* 1987. Modulation of human natural killer cell cytotoxic activity, lymphokine production, and interleukin 2 receptor expression by thymic hormones. *J. Immunol.* **139:** 2338–2343.
8. Sztein, M., S. Serrate & A. Goldstein. 1986. Modulation of interleukin-2 receptor expression on normal human lymphocytes by thymic hormones. *Proc. Natl. Acad. Sci. USA* **83:** 6107–6111.
9. Sztein, M. & S. Serrate. 1989. Characterization of the immunoregulatory properties of thymosin alpha 1 on interleukin-2 production and interleukin-2 receptor expression in normal human lymphocytes. *Int. J. Immunopharmacol.* **11:** 789–800.
10. Leichtling, K.D., S.A. Serrate & M.B. Sztein. 1990. Thymosin alpha 1 modulates the expression of high affinity interleukin-2 receptors on normal human lymphocytes. *Int. J. Immunopharmacol.* **12:** 19–29.
11. Svedersky, L., A. Hui, L. May, *et al.* 1982. Induction and augmentation of mitogen-induced immune interferon production in human peripheral blood lymphocytes by Na-desacetylthymosin alpha 1. *Eur. J. Immunol.* **12:** 244–247.
12. Hsia, J., N. Sarin, J.H. Oliver & A.L. Goldstein. 1989. Aspirin and thymosin increase interleukin-2 and interferon-gamma production by human peripheral blood lymphocytes. *Immunopharmacology* **17:** 167–173.
13. Mutchnick, M.G., H.D. Appelman, H.T. Chung, *et al.* 1991. Thymosin treatment of chronic hepatitis B: a placebo-controlled pilot trial. *Hepatology* **14:** 409–415.
14. Ershler, W., J. Hebert, A. Blow, *et al.* 1985. Effect of thymosin alpha one on specific antibody response and susceptibility to infection in young and aged mice. *Int. J. Immunopharmacol.* **7:** 465–471.
15. Effros, R.B., A. Casillas & R.L. Walford. 1988. The effect of thymosin alpha 1 on immunity to influenza in aged mice. In *Aging: Immunology and Infectious Disease.* **Vol. 1.** 31–40. Mary Ann Liebert, Inc. New York, NY.
16. Di Francesco, P., F. Pica, S. Marini, *et al.* 1992. Thymosin alpha one restores murine T-cell-mediated responses inhibited by *in vivo* cocaine administration. *Int. J. Immunopharmacol.* **14:** 1–9.
17. Ershler, W., A. Moore & M. Socinski. 1984. Influenza and aging: age-related changes and the effects of thymosin on the antibody response to influenza vaccine. *J. Clin. Immunol.* **4:** 445–454.

18. Ciancio, A., P. Andreone, S. Kaiser, *et al.* 2012. Thymosin alpha 1 with peginterferon alfa-2a/ribavirin for chronic hepatitis C not responsive to IFN/ribavirin: an adjuvant role? *J. Viral Hepat.* **19**(Suppl. 1): 52–59.
19. Zhang, P., J. Chan, A.M. Dragoi, *et al.* 2005. Activation of IKK by thymosin alpha1 requires the TRAF6 signalling pathway. *EMBO Rep.* **6:** 531–537.
20. Arrighi, J.F., M. Rebsamen, F. Rousset, *et al.* 2001. A critical role for p38 mitogen-activated protein kinase in the maturation of human blood-derived dendritic cells induced by lipopolysacharide, TNF-q, and contact sensitizers. *J. Immunol.* **166:** 3837–3845.
21. Iijima, N., Y. Yanagawa & K. Onoe. 2003. Role of early- or late-phase activation of p38 mitogen-activated protein kinase induced by tumor necrosis factor-a or 2,4-dinitrochlorobenzene during maturation of murine dendritic cells. *Immunology* **110:** 322–328.
22. Yoshimura, S., J. Bondeson, F.M. Brennan, *et al.* 2001. Role of NFkappaB in antigen presentation and development of regulatory T cells elucidated by treatment of dendritic cells with proteasome inhibitor PSI. *Eur. J. Immunol.* **31:** 1883–1893.
23. Knutsen, A.P., J.J. Freeman, K.R. Mueller, *et al.* 1999. Thymosin-alpha1 stimulates maturation of CD34+ stem cells into CD3 + 4+ cells in an *in vitro* thymic epithelia organ coculture model. *Int. J. Immunopharmacol.* **21:** 15–26.
24. Pierliugi, B., C. D'Angelo, F. Fallarino, *et al.* 2010. Thymosin alpha 1: the regulator of regulators? *Ann. N.Y. Acad. Sci.* **1194:** 1–5.
25. Shrivastava, P., S.M. Singh & N. Singh. 2004. Effect of thymosin alpha 1 on the antitumor activity of tumor-associated macrophage-derived dendritic cells. *J. Biomed. Sci.* **11:** 623–630.
26. Shen, S., J. Josselson, C. McRoy, *et al.* 1987. Effect of thymosin alpha 1 on heptavax-B vaccination among hemodialysis patients. *Kidney Int.* **31:** 217.
27. Schulof, R.S., M.J. Lloyd, P.A. Cleary, *et al.* 1985. A randomized trial to evaluate the immunorestorative properties of synthetic thymosin alpha 1 in patients with lung cancer. *J, Biol. Response Modifiers* **4:** 147–158.
28. Stefanini, G.F., F.G. Foschi, E. Castelli, *et al.* 1998. Alpha-1-thymosin and transcatheter arterial chemoembolization in hepatocellular carcinoma patients: a preliminary experience. *Hepatogastroenterology* **45:** 209–215.
29. de Weck, A., F. Kristensen, F. Joncourt, *et al.* 1984. Lymphocyte proliferation, lymphokine production, and lymphocyte receptors in ageing and various clinical conditions. *Springer Semin. Immunopathol.* **7:** 273 289.
30. Makinodan, T. & W. Peterson. 1962. Relative antibody-forming capacity of spleen cells as a function ofage. *Proc. Natl. Acad. Sci. USA* **48:** 234–238.
31. Wade, A.W. & M.R. Szewczuk. 1984. Aging, idiotype repertoire shifts, and compartmentalization of the mucosal-associated lymphoid system. *Adv. Immunol.* **36:** 143–188.
32. Ershler, W., A. Moore, M. Hacker, *et al.* 1984. Specific antibody synthesis *in vitro*. II. Age-associated thymosin enhancement of antitetanus antibody synthesis. *Immunopharmacology* **8:** 69–77.
33. Bramwell, S.P., D.J. Tsakiris, J.D. Briggs, *et al.* 1985. Dinitrochlorobenzene skin testing predicts response to hepatitis B vaccine in dialysis patients. *Lancet* **1:** 1412–1415.
34. Dammin, G., N. Couch & J. Murray. 1957. Prolonged survival of skin homografts in uremic patients. *Paper presented at: Second Tissue Homotransplantation Conference.* New York, NY.
35. Lawrence, H.S. 1965. Uremia: nature's immunosuppressive device [editorial]. *Ann. Intern. Med.* **62:** 166–170.
36. Revie, D., S. Shen, J. Ordonez, *et al.* 1985. T-cell subsets and status of hepatitis B surface antigen and antibody in end-stage renal disease patients. *Kidney Int.* **27:** 150.
37. Sanders, C.V., Jr., L.P. Luby, J.P. Sanford & A.R. Hull. 1971. Suppression of interferon response in lymphocytes from patients with uremia. *J. Lab. Clin. Med.* **77:** 768–776.
38. Stevens, C.E., H.J. Alter, P.E. Taylor, *et al.* 1984. Hepatitis B vaccine in patients receiving hemodialysis: immunogenicity and efficacy. *N. Engl. J. Med.* **311:** 496–501.
39. Grob, P., U. Binswanger, K. Zaruba, *et al.* 1983. Immunogenicity of a hepatitis B subunit vaccine in hemodialysis and in renal transplant recipients. *Antiviral Res.* **3:** 43–52.
40. Crosnier, J., P. Jungers, A.M. Courouc, *et al.* 1981. Randomised placebo-controlled trial of hepatitis B surface antigen vaccine in French haemodialysis units: I, medical staff. *Lancet* **1:** 455–459.
41. Gravenstein, S., W.B. Ershler, S. Drumaskin, *et al.* 1986. Anti-influenza antibody response: augmentation in elderly –non-responders by thymosin alpha 1. *The Gerontologist* **26:** 150A.
42. Gravenstein, S., E.H. Duthie, B.A. Miller, *et al.* 1989. Augmentation of influenza antibody response in elderly men by thymosin alpha one. A double-blind placebo-controlled clinical study. *J. Am. Geriatr. Soc.* **37:** 1–8.
43. McConnell, L., S. Gravenstein, E. Roecker, *et al.* 1989. Augmentation of influenza antibody levels and reduction in attack rates in elderly subjects by thymosin alpha 1. *The Gerentologist* **29:** 188A.
44. Shen, S.Y., Q.B. Corteza, J. Josselson, *et al.* 1990. Age-dependent enhancement of influenza vaccine responses by thymosin in chronic hemodialysis patients. In *Biomedical Advances in Aging*. A.L. Goldstein, Ed.: 523–530. Plenum Press. New York, NY.
45. Shen, S., J. Josselson, C. McRoy, *et al.* 1987. Effects of thymosin alpha 1 on peripheral T-cell and Heptavax-B vaccination in previously non-responsive hemodialysis patients. *Hepatology* **7:** 1120.
46. Tuthill, C., A. DeRosa, R. Camerini, *et al.* 2010. The immunomodulatory peptide thymosin alpha 1 enhances response to influenza vaccine. *National Foundation for Infectious Diseases 13th Annual Meeting*. Bethesda, MD.
47. Carraro, G., A. Naso, E. Montomoli, *et al.* 2012. Thymosin alpha 1 (Zadaxin) enhances the immunogenicity of an adjuvanted pandemic H1N1v influenza vaccine (Focetria) in hemodialyzed patients: a pilot study. *Vaccine* **30:** 1170–1180.

Ann. N.Y. Acad. Sci. ISSN 0077-8923

ANNALS OF THE NEW YORK ACADEMY OF SCIENCES
Issue: *Thymosins in Health and Disease*

The use of angiogenic-antimicrobial agents in experimental wounds in animals: problems and solutions

Paritosh Suman,[1] Harikrishnan Ramachandran,[1] Sossy Sahakian,[1] Kamraan Z. Gill,[2] Basil A. J. Horst,[2] Shanta M. Modak,[1] and Mark A. Hardy[1]

[1]Department of Surgery, [2]Department of Pathology, Columbia University, New York, New York

Address for correspondence: Paritosh Suman, Columbia University, Surgery, 630 W 168th St., P & S 17-501, New York, NY 10032. suman.paritosh@gmail.com

A topical combination (silvathymosin) of natural proangiogeneic protein thymosin β4 (Tβ4) and antimicrobial silver sulfadiazine was hypothesized to promote the healing of large, full-thickness, clean or infected wounds in rats. Silvathymosin showed the fastest wound healing (85%) followed by silver sulfadiazine (84%) and Tβ4 (72%). In the infected groups, the healing pattern was different, as Tβ4 and silvathymosin groups did not show similar wound healing. Wound histopathology and VEGF and KI67 immunohistochemical assessment of angiogenesis was consistent and correlated well with the tempo of healing of the acute wounds. These preliminary data demonstrate the more rapid acute wound healing properties of the combination formulation of thymosin β4 and silver sulfadiazine as compared to these agents alone. This novel agent could prove an effective treatment modality for debilitating chronic wounds and decubitus ulcers.

Keywords: chronic wound healing; thymosin β4; animal models

Introduction

The nonhealing chronic ulcer, usually due to pressure, venous stasis, diabetic neuropathy, and other ischemic conditions, is a major health problem that takes a significant toll on patients, especially the elderly. An estimated annual health care expenditure for chronic ulcer–related care is $5 to 10 billion dollars.[1] Beyond the extraordinarily high economic expenses, these nonhealing wounds contribute to persistent pain and loss of gainful employment in vulnerable elders.[2] Pressure and venous stasis ulcers result in nonhealing wounds in approximately 1.5 to 3 million[3] and a half million[4] patients every year, respectively, while ischemic and neuropathic diabetic ulcers lead to 85,000 lower limb amputations.[5] Unfortunately, there has been no significant breakthrough in the management paradigms of chronic nonhealing wounds in the recent past, despite the dedicated efforts of the medical and scientific community.

Pathophysiology of wound healing

The standard description of normal wound healing involves several orderly and well-orchestrated steps, which include (1) initial inflammation and the removal of the destroyed or injured parts; (2) cellular proliferation, which contributes to both the "clean up" process and the start of reconstruction of the tissues; (3) angiogenesis, which creates the "roads (capillary network)" that permit the arrival of new cells, the removal of debris, and provisions of nutrition, minerals, oxygen, vitamins, and other needed materials for wound healing and remodeling; (4) migration of fibroblast and keratinocyte, the cells critical to the repair of the wound; and (5) wound remodeling (collagen pattern alteration).

Disruption of this orderly and temporal series of cellular events results in a structurally flawed and dysfunctional tissue repair, which is the hallmark of a chronic wound.[6] Characteristically, these chronic wounds display prolonged inflammation and

doi: 10.1111/j.1749-6632.2012.06653.x

dysregulated cellular defense mechanisms and also frequently harbor microbial superinfection.[7,8] A duration of six to eight weeks of nonhealing usually leads to the classification of such wounds as chronic.[7,9]

Although there are diverse etiologies for different types of chronic wounds, these ulcers share a common pathway, which leads to the development of chronic wounds in humans. Ischemia and tissue hypoxia seem to be the most common final pathways to chronicity. Cellular replicative senescence has also been proposed to be responsible for delayed wound healing.[10] Interestingly, surgical debridement can convert a chronic wound to an acute wound by removal of senescent wound cells through debridement. This can potentially explain the effectiveness of surgical interventions. Pathophysiology of chronic pressure (decubiti) ulcers is based, usually on several interrelated factors, including altered body pressure distribution, malnutrition, and frequently a concomitant infectious process with tissue ischemia.[11–14] In the case of venous stasis ulcers, incompetency of the deep, superficial, and perforator veins leads to venous hypertension resulting in hemodynamic changes in the microcirculation in the most dependent area, that is, the ankle area. This results in increased capillary pressure, edema, and, finally, skin ulceration.[15] Compression bandaging or compression hose are the treatments of choice for venous ulcers. Diabetic ulcers are thought to develop in the presence of ischemia from occlusive vascular disease of both the large and small blood vessels of the extremities in association with peripheral neuropathy and loss of sensation. Repetitive extremity trauma[16] in the face of ischemia and loss of sensation are considered the main determinants for the development of a diabetic ulcer. Impaired leukocyte function[17] and other immune deficiencies predispose the diabetic to infectious complications of these ulcers. Glycemic control, debridement, treatment of infection, and off-loading are the cornerstones for the treatment of diabetic foot ulcers. Pressure ulcers develop when the pressure on the tissue occludes the circulation, resulting in tissue damage and death. This ischemia due to pressure is worsened by the frequent resumption of circulation with repositioning of the patient. This ischemia–reperfusion syndrome can be especially damaging to the deep tissues as well as the skin. Treatment of pressure ulcers, especially ones on the trunk, requires debridement, constant off-loading of pressure by frequent repositioning and special support surfaces, nutritional support, and prevention of fecal and urinary contamination. In the final analysis, all chronic ulcers have one common problem and that is local ischemia of the tissues causing tissue damage. All treatments for these ulcers depend on restoring adequate blood supply to the affected area and regeneration of the damaged microcirculation and matrix.

Thymosin β4

Thymosin β4 (Tβ4) is a naturally occurring, highly conserved, water-soluble protein, consisting of 43 amino acids that are present in all eukaryotic cells, especially in human circulating cells and platelets.[18] Tβ4 promotes cell morphogenesis and motility by regulating the dynamics of the actin cytoskeleton.[19] Over the past 15 years, studies have implicated Tβ4 in a number of cellular events, such as angiogenesis,[20,21] wound healing,[22,23] hair growth,[24] apoptosis,[25,26] and inflammatory responses.[26] Specifically, as a promoter of tissue repair and regeneration, Tβ4 augments angiogenesis, cell migration, reepithelialization, and downregulates inflammation among other effects[18,21,22,26–33] (Table 1). Interestingly, the functions of Tβ4 differ significantly from other growth factors. Unlike the latter, Tβ4 acts on a wide array of cells and has a wide range of biological effects that include antiapoptosis, antifibrotic, and antimicrobial activities,[34–36] all of which have a potential favorable role in the healing of chronic wounds. The various physiological and biochemical properties of Tβ4 suggest an effective role for Tβ4 as a primary or as a supportive agent for healing of chronic wounds when compared to growth factors.

Thymosin β4 and wound healing

As a multifactorial protein, Tβ4 stimulates multiple stages of the wound repair process in a variety of tissues. Specifically in wound healing of skin, Tβ4 has been shown to increase wound closure by promoting keratinocyte migration, collagen deposition, and angiogenesis. Evidently, Tβ4 has the potential to be an important healing agent for chronic wounds, as is suggested in Phase II/III clinical studies described elsewhere in this issue by T. Treadwell *et al.*

Silver sulfadiazine and wound healing

Topical antimicrobial silver sulfadiazine (silvadene) is the standard of care in the treatment of burns,

Table 1. Major wound healing properties of thymosin β4

Physiological property	Cellular mechanisms	Significance in wound healing	References
Angiogenesis	Increases endothelial cell migration and tubule formation	Promotes wound healing by optimizing delivery of nutrients and oxygen	20, 21
Antiapoptotic/ prosurvival	Downregulates post-injury cellular apoptosis	Limits death and damage to tissue under ischemic/metabolic stress	25, 26
Antifibrotic	Inhibits excessive collagen deposition	Optimal scarring	36
Anti-inflammatory	Reduces inflammatory cell population and downregulates the inflammatory pathway	Controls unabated wound healing response, promotes orderly wound healing especially in corneal wounds	26, 27
Antimicrobial	Prevents and reduces bacterial colonization and superinfection	Reduced need for essential nutrients and oxygen for wound healing	35
Cell migration	Inhibits neutrophil chemotaxis *in vitro,* stimulates endothelial cell, keratinocyte, and fibroblast migration	Ocular and dermal wounds	22, 26
Stem cell proliferation	In dermis	Accelerated cutaneous wound healing	23

especially in partial thickness burn wounds, because of its ability to prevent opportunistic infection and to reduce the bacterial load. In 1974, Fox and Modak[37] demonstrated the efficacy of silvadene in the treatment of burn wounds, and others have used it to treat chronic wounds as well. Application of silver sulfadiazine for chronic nonhealing wounds could augment the healing properties of other topical agents by creating a favorable microenvironment free of pathogens. Of note is that the wide spectrum of antimicrobial efficacy, ease of application, and availability make silvadene the drug of choice in the treatment of chronic ulcers.

Clinical studies

Phase II clinical trials on the use of Tβ4 for chronic pressure and venous stasis ulcers (see Ref. 61) have been completed, and though not conclusive, show an impressive positive trend for acceleration of healing of both kinds of wounds. Of interest is the observation that a 0.02% formulation of Tβ4 showed a better wound closure response compared to higher and lower concentrations, as well as compared to the placebo. Tβ4 achieved a wound size reduction rate of 80% and 100% for chronic pressure and venous stasis ulcers, respectively. Similar encouraging results were found in a European multicenter phase II trial on the use of Tβ4 in venous stasis ulcers.[38] To the best of our knowledge there are no clinical studies that have demonstrated the efficacy of Tβ4 in human diabetic wounds.

Many attempts at accelerating the healing of venous stasis ulcers have so far been tried with limited success. The most common treatment involves the use of compression bandaging in conjunction with wound therapies. Treatment of the ulcer entails debridement of necrotic tissue, treatment of infection and/or biofilm, control of exudate, and the use of advanced therapies such as growth factor therapy and tissue engineered skin products.[39] In a recent comprehensive literature review, Perrin *et al.*[40] evaluated the use of venoactive drugs in the treatment of ulcers due to chronic venous hypertension. Only one product, micronized purified flavonoid fraction, had any effect on the healing of venous ulcers. A robust treatment for venous stasis ulcers has not been established.

Animal models for wound healing

Since animal models remain the initial mainstay for the study of chronic wound-healing mechanisms, the lack of such experimental models has hindered progress in the clinical area. Presentation of chronic wounds in humans is, by its nature, delayed, while the experimental studies of wounds have usually focused on acute injuries of various etiologies. It is our intention in this study to discuss both the pitfalls of studies of chronic wound healing and the available solutions to such problems, using appropriate animal models to study early pathogenesis and triggering mechanisms of such wounds. Investment of extensive resources and research efforts in studying wound healing have so far not been able to overcome the discordance between animal models and human tissue pathophysiology.[41] Development of novel animal models may solve this problem in the near future (Table 2).

Depending on the depth of the ulcer, starting with the skin and sometimes extending to the underlying tissue up to the bone, pressure (decubiti) ulcers are divided into stages I through IV. Most of the described methods of induction of pressure ulcers in animal models involve only skin and subcutaneous adipose tissue.[42–44] Although intermittent reperfusion has been shown to be essential for the development of pressure ulcers,[45,46] the existing experimental models are subject to constant ischemia with no tissue reperfusion. Reperfusion could be important in the development of pressure ulcers, as is evident in hospitalized patients and nursing home residents where, despite the implementation of frequent repositioning, occurrence of decubitus ulcers is fairly common.

Similar problems are faced by the studies of various models of chronic and acute venous hypertension.[47] These include acute venous occlusion, creation of arteriovenous fistula,[48,49] and ligation of

Table 2. Wound-healing animal models and their deficiencies

Animal model and wound type	Findings and deficiencies	References
Pressure ulcers		
Skin and subcutaneous adipose tissue	Lack of intermittent perfusion	42–44
Venous ulcers		
Acute venous occlusion	Failure to produce chronic venous ulcers	47
Creation of arteriovenous fistula		48, 49
Ligation of limb veins		50
Diabetic ulcers		
db/db genetically diabetic animals	Wound reepithelialization, slow wound healing	32
With thymosin β4	thymosin β4 accelerates wound size reduction	32
Thymosin β4 and nondiabetic cutaneous wounds	Failure to create chronic wound conditions	22, 32
Porcine skin wound healing model	Close resemblance to human skin; high incidence of fulminant infection and sepsis	51–55
Small mammal cutaneous wound healing models	Ease of handling and low cost; wound heals by contraction not reepithelialization	
Interesting animal wound healing models		
Partial thickness cutaneous wound between apposing magnetic and metallic plates	Intermittent ischemia/perfusion; only partial thickness wound	42
Full thickness stage IV wounds between apposing magnetic and metallic plates	Intermittent ischemia/perfusion	58
Full thickness wound in raised skin flaps	Chronic ischemia, porcine model mimicking human wounds	59
Spinal cord transection to create full thickness stage IV skin ulcers	Mimics human decubiti ulcers; small mammal model	60

veins of large limbs.[50] Although such experimental models have contributed substantially to the understanding of pathophysiology of chronic venous hypertension, they have not narrowed the gap between animal research of chronic venous hypertension and the clinical findings in humans, as none of these models resulted in spontaneous formation of chronic venous ulcers.

Studies of cutaneous diabetic ulcers have mostly been conducted in genetically diabetic db/db mice. The *db/db* mouse has a genetic mutation in the leptin receptor (LEPR), which renders these mice type 2 diabetic and obese with peripheral neuropathy. Peripheral neuropathy and wound reepithelialization (instead of contraction), and comparatively slower wound healing makes *db/db* mice an attractive model to study cutaneous nonhealing wounds. Other rodent models of diabetes, including the use of streprozotocin-induced diabetes, have not been useful for wound studies. The study by Philip *et al.*[32] on the use of Tβ4 in diabetic animal wounds reported a significant increase in the collagen deposition and wound size reduction in *db/db* diabetic mice compared to wild-type controls. The study was limited to seven days with all relatively small wounds showing rapid healing response. Similarly, the majority of studies pertaining to the healing of nondiabetic cutaneous wounds using topical application of Tβ4 were performed in models that failed to create chronic wound healing conditions.[22,32] The present study did not avoid this problem. The effectiveness of Tβ4 in chronic wounds in appropriate experimental models remains to be investigated.

In our experimental rat model, we surgically created large 3×3 cm full-thickness dorsal skin wounds with the hope of eliciting a chronic wound healing response. We avoided a small wound by using a punch, as used by other investigators in the past.[32] The animals were divided into four groups of six rats each: (1) the control group, (2) group treated with 0.01% Tβ4 gel, (3) group treated with 1% silver sulfadiazine, and (4) group treated with a combination agent Tβ4 + silver sulfadiazine (silvathymosin). Formulations under study were applied daily with daily dressing change of all surgical wounds. Wounds were measured on day 0, day 7, and day 14, with determination of relative wound contraction. All three formulations resulted in faster early wound healing than the untreated control.

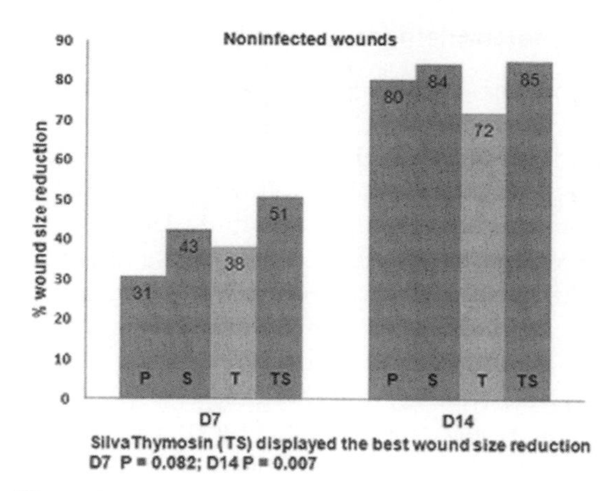

Figure 1. In noninfected rat skin–wound experiments, combination silvathymosin (TS) showed the best wound size reduction on D7 ($P = 0.082$) and D14 ($P = 0.007$) healing response. D, day; P, control; S, silver sulfadiazine (silvadene); T, Tβ4; TS, silver sulfadiazine + thymosin β4 (Silvathymosin).

The combination agent silvathymosin showed the fastest (Fig. 1, D7 $P = 0.007$) wound contraction in comparison to Tβ4 or silvadene alone. Wound healing response of silvathymosin was significant at day 7 ($P = 0.04$) and day 14 ($P = 0.018$) when compared to Tβ4. Histological analysis supported the wound contraction observation where silvathymosin treated dermal wounds were found to have the most abundant granulation tissues and deposition of collagen. Immunohistochemical staining of tissues from wounds treated with silvathymosin for Ki-67 and VEGF antigens also revealed the most effective cellular proliferation and angiogenesis in this group. These results confirmed the known wound-healing properties of Tβ4 and silver sulfadiazine.

A separate series of similarly designed experiments were performed in four other groups of rats to assess the wound healing properties of Tβ4 and silvathymosin in infected cutaneous wounds. In the latter study, wounds made in the same fashion as previously described were infected with *Staphylococcus aureus* (10^5 bacteria). The animals were treated with the topical agents in an identical manner to that previously described in the first series of experiments. In addition to the observation and biopsy protocol that we described for the first series of animals, we also obtained semiquantitative bacterial estimation on each wound with swab cultures and determined quantitative bacterial loads by serial wound biopsies. The silvathymosin group demonstrated statistically significantly lower bacterial load

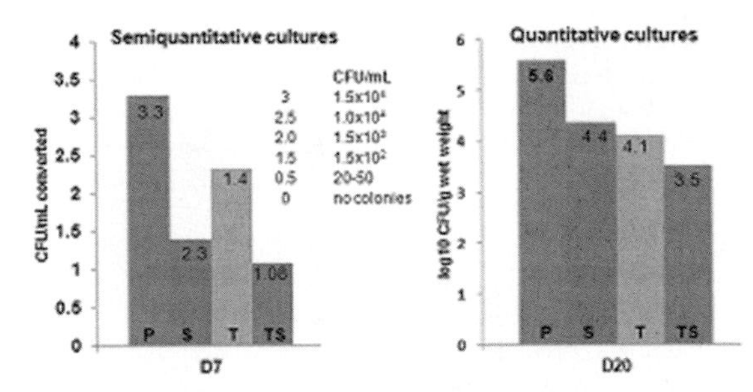

Figure 2. In infected rat skin–wound experiments, combination silvathymosin (TS) had the lowest bacterial load in semiquantitative (D7 $P = 0.016$) and quantitative (D20 $P = 0.004$) cultures. D, day; P, control; S, silver sulfadiazine (silvadene); T, thymosin β4 (Tβ4); TS, silver sulfadiazine + thymosin β4 (silvathymosin).

(Fig. 2, semiquantitative D7 $P = 0.016$; quantitative $P = 0.004$) when compared to silvadene or Tβ4 alone. Contrary to the good antimicrobial results, surprisingly, the wound-size reduction was slower in the silvathymosin and Tβ4-treated wounds in comparison to the wounds treated with silvadene alone. Since the silvathymosin used in this experiment was 90 days old, it is possible that the formulation was unstable, and this might have played a role in the observed lack of wound healing. Overall, in our experimental cutaneous wound-healing model, Tβ4 and silvathymosin demonstrated significant early wound healing.

As ischemia is thought to be a critical component in the development of a chronic wound, the animal models used for this purpose depend on the creation of ischemic conditions in one form or another, which we did not achieve in our original series of experiments. In addition to the reduced vascular supply, local tissue conditions also determine the extent of tissue hypoxia in the microenvironment of a chronic wound; thus, global ischemia does not always correlate with effective tissue hypoxia. The majority of the experiments on wound healing have been conducted on small mammals whose anatomy and wound healing characteristics differ significantly from that of human skin. Cutaneous wounds in small mammals heal mostly by contraction rather than by reepithelialization as they do in humans. To achieve reepithelialization similar to human skin, large animal wound models may be more reliable. Large animal models, such as the pig, whose skin best resembles human skin,[51–54] unfortunately are subject to an unusually high incidence of uncontrolled infection and sepsis as a result of wound exudate collections following prolonged dressing conditions. Porcine skin resembles human skin anatomically and also demonstrates very similar physiological properties including response to the various growth factors.[51,55] Despite the advantages of porcine cutaneous wound models, rat and mouse models remain the most frequently used due to their low cost and ease of handling.

Our findings in the rat skin wound model were consistent with previous observations on the use of topical Tβ4 in wound healing. Without replicating appropriate conditions for chronic wound healing, we observed an accelerated wound-healing response in groups treated with either Tβ4 alone or its combination with silver sulfadiazine. Failure to maintain constant tissue ischemia, and our inability to recreate human chronic pressure or the diabetic ulcer milieu, were the major deficiencies in our pilot experiments using the simplest, but not the most appropriate, animal model.

Evidently, the major hindrance to the development of an ideal animal model of dermal wound healing is the initiation and maintenance of a delayed and dysregulated chronic wound–healing response that would closely mimic human chronic wound conditions.[56] Most animal models of wound healing, including our experimental rat model, exhibit the shortcoming of rapid and aggressive wound healing.[57] Pierce *et al.* developed an ingenious method to create pressure ulcers in rats by generating intermittent pressure ischemia/reperfusion cycles across the skin between an embedded steel plate and an overlying magnet.[42] Wassermann[58] reported a modification of this method by implanting a submuscular steel plate and thus was able to create a full thickness stage IV pressure ulcer. Another excellent chronic ishemic cutaneous wound model was also recently described by Roy *et al.*[59] In their study, porcine skin flaps were raised to create an ischemic microenvironment. A full-thickness wound placed in these ischemic skin flaps would be expected to mimic chronic wound healing pathways. Maintenance of extended cutaneous ischemia, direct functional measurement of tissue oxygenation and blood flow by skin perfusion pressure (SPP), and a simultaneous comparison to pair-matched nonischemic cutaneous wounds make this model an elegant substitute for studies of chronic wound conditions in humans. Another model that deserves consideration is that described by Lin *et al.*[60] who used a spinal cord transection technique to recreate a deep tissue injury pressure ulcer model in experimental rats that closely reflected the human pressure ulcer conditions.

In summary, an ideal animal model of chronic cutaneous wound healing must satisfy certain pathophysiological and experimental criteria, which were not achieved in our model. The most important factor is the creation of long-term ischemic tissue microenvironment. The cutaneous wound should be of full thickness and large enough to allow healing by reepithelialization with the provision of repetitive cycles of ischemia and reperfusion. An objective measurement of SPP for tissue hypoxia must be an integral part of the experimental model. Additionally, the experimental protocol must include the provision of frequent dressing changes to prevent severe sepsis, while maintaining a biofilm of chronic bacterial superinfection. The animal model should be reproducible and flexible to allow the modification of testing variable conditions, which need to mimic those found in various types of wounds being studied, for example, diabetic, chronic venous ulcer, or pressure ulcers.

Conclusions

Major efforts and funding are being invested in the development of novel therapeutic strategies to effectively treat chronic wounds. Innovative and promising experimental models are available for use to achieve this goal. Unfortunately, the efforts of the scientific community remain fragmented. It appears clear that a single experimental model or therapy will be insufficient to develop a solution(s) to the vexing problems posed by various types of chronic wounds. There is a distinct need to adopt a collaborative multidisciplinary approach by investigators in various fields of study whose valuable resources and ideas could be complementary and be freely shared. A focused joint research effort offers a promising approach to helping millions of patients suffering nonhealing chronic wounds of various etiologies.

Conflicts of interest

The authors declare no conflicts of interest.

References

1. Kuehn, B.M. 2007. Chronic wound care guidelines issued. *JAMA* **297:** 938–939.
2. Vileikyte, L. 2001. Diabetic foot ulcers: a quality of life issue. *Diab. Metab. Res. Rev.* **17:** 246–249.
3. Lobmann, R., A. Ambrosch & G. Schultz. 2002. Expression of matrix-metalloproteinases and their inhibitors in the wounds of diabetic and nondiabetic patients. *Diabetologia* **45:** 1011–1016.
4. Olin, J.W., K.M. Beusterien & M.B. Childs. 1999. Medical costs of treating venous stasis ulcers: evidence from a retrospective cohort study. *Vasc. Med.* **4:** 1–7.
5. Kantor, J. & D.J. Margolis. 2001. Treatment options for diabetic neuropathic foot ulcers: a cost-effectiveness analysis. *Dermatol. Surg.* **27:** 347–351.
6. Papadopoulos, E., S. Leibenhaut & C.N. Durfor. 2006. Guidance for industry: chronic cutaneous ulcer and burn wounds – developing products for treatment. U.S. Department of Health and Human Services, Food and Drug Administration.
7. Tomic-Canic, M., M.S. Agren & O.M. Alvares. 2004. Epidermal repair and chronic wounds. In *The Epidermis in Wound Healing*. D.T. Rovie & H. Maibach, Eds.: 25–57. CRS Press. New York.
8. Bowler, P.G. 2002. Wound pathophysiology, infection and therapeutic options. *Ann. Med.* **34:** 419.
9. Fonder, M.A., G.S. Lazarus & D.A. Cowan. 2008. Treating the chronic wound: a practical approach to the care of nonhealing wounds and wound care dressings. *J. Am. Acad. Dermatol.* **58:** 185–206.
10. Wright, W.E. & J.W. Shay. 1991. The two-stage mechanism controlling cellular senescence and immortalization. *Exp. Gerontol.* **27:** 383–389.
11. Bennett, L., D. Kavner & B.K. Lee. 1979. Shear vs pressure as causative factors in skin blood flow occlusion. *Arch. Phys. Med. Rehabil.* **60:** 309–314.
12. Meijer, J.H., P.H. Germs & H. Schneider. 1994. Susceptibility to decubitus ulcer formation. *Arch. Phys. Med. Rehabil.* **75:** 318–323.
13. Goldstein, B. & J. Sanders. 1998. Skin response to repetitive mechanical stress: a new experimental model in pig. *Arch. Phys. Med. Rehabil.* **79:** 265–272.
14. Whitney, J., L. Phillips & R. Aslam. 2006. Guidelines for the treatment of pressure ulcers. *Wound Repair Regen.* **14:** 663–679.
15. Kerstein, M.D., & V. Gahtan. 1998. Outcomes of venous ulcer care: results of longitudinal study. *Ostomy/Wound Manage.* **44:** 52–64.
16. Brem, H., P. Sheehan & A.J. Boulton. 2004. Protocol for treatment of diabetic foot ulcers. *Am. J. Surg.* **187:** 1S–10S.
17. Delamaire, M., D. Maugendre & M. Moreno. 1997. Impaired leucocyte functions in diabetic patients. *Diabet. Med.* **14:** 29–34.
18. Goldstein, A.L., E. Hannappel & H.K. Kleinman. 2005. Thymosin beta 4: actin-sequestering protein moonlights to repair injured tissues. *Trends. Mol. Med.* 11: 421–429.
19. Huff, T., C.S.G. Muller & A.M. Otto. 2001. beta -Thymosins small acidic peptides with multiple functions. *Int. J. Biochem. Cell Biol.* **33:** 205–220.
20. Grant, D.S., W. Rose & C. Yaen. 1999. Thymosin beta 4 enhances endothelial cell differentiation and angiogenesis. *Angiogenesis* **3:** 125–135.

21. Malinda, K.M., A.L. Goldstein & H.K. Kleinman. 1997. Thymosin beta 4 stimulates directional migration of human umbilical vein endothelial cells. *FASEB J.* **11:** 474–481.
22. Malinda K.M., G.S. Sidhu & H. Mani. 1999. Thymosin beta 4 accelerates wound healing. *J. Invest. Dermatol.* **113:** 364–368.
23. Philp, D., A.L. Goldstein & H.K. Kleinman. 2004. Thymosin beta 4 promotes angiogenesis, wound healing and hair follicle development. *Mech. Ageing Dev.* **125:** 113–115.
24. Philp, D., S. St-Surin & H.J. Cha. 2007. Thymosin beta 4 induces hair growth via stem cell migration and differentiation. *Ann. N. Y. Acad. Sci.* **1112:** 95–103.
25. Smart, N., C.A. Riseboro & A.A. Melville. 2007. Thymosin beta 4 induces adult cardiac progenitor mobilization and neovascularization. *Nature* **445:** 177–182.
26. Young, J.D., A.J. Lawrence & A.G. Mac Lean. 1999. Thymosin beta 4 sulfoxide is an anti-inflammatory agent generated by monocytes in the presence of glucocorticoids. *Nat. Med.* **5:** 1424–1427.
27. Sosne, G., P. Qiu & P.L. Christopherson. 2007. Thymosin beta 4 suppression of corneal NFkappaB: a potential anti-inflammatory pathway. *Exp. Eye Res.* **84:** 663–669.
28. Mannherz, H.G. & E. Hannappel. The beta thymosins: intracellular and extracellular activities of a versatile actin binding protein family. *Cell Motil. Cytoskeleton* **66:** 839–851.
29. Roy, P., Z. Raifur & D. Jones. Local photorelease of caged thymosin beta 4 in locomoting keratocytes causes cell turning. *J. Cell Biol.* **153:** 1035–1048.
30. Cha, H.J., M.J. Jeong & H.K. Kleinman. 2003. Role of thymosin beta 4 in tumor metastasis and angiogenesis. *J. Nat. Cancer Inst.* **95:** 1674–1680.
31. Grant, D.S., J.L. Kinsella & M.C. Kibbey. A novel role for thymosin beta 4: a Matrigel-induced gene involved in endothelial cell differentiation and angiogenesis. *J. Cell Sci.* **108:** 3685–3694.
32. Philp, D., M. Badamchian & B. Scheremeta. 2003. Thymosin beta 4 and a synthetic peptide containing its actin-binding domain promote dermal wound repair in db/db diabetic mice and in aged mice. *Wound Repair Regen.* **11:** 19–24.
33. Badamchian, M., M.O. Fagarasan & R.L. Danne. 2003. Thymosin beta 4 reduces lethality and downregulates inflam matory mediators in endotoxin-induced septic shock. *Int. Immunopharmacol.* **3:** 1225–1233.
34. Popoli, P., R. Pepponi & A. Martire. Neuroprotective effects of thymosin beta 4 in experimental models of excitotoxicity. *Ann. N. Y. Acad. Sci.* **1112:** 219–224.
35. Huang, L.C., D. Jean & R.J. Proske. 2007. Ocular surface expression and in vitro activity of antimicrobial peptides. *Current. Eye Res.* **32:** 595–609.
36. Cavasin, M.A. 2006. Therapeutic potential of thymosin beta 4 and its derivative N-acetyl-seryl-aspartyl-lysylproline (Ac-SDKP) in cardiac healing after infarction. *Am. J. Cardiovasc. Drugs* **6:** 305–311.
37. Fox, C.L. Jr. & S.M. Modak. 1974. Mechanism of silver sulfadiazine action on burn wound infections. *Antimicrob. Agents Chemother.* **5:** 582–588.
38. Guarnera, G., A. DeRosa & R. Camerini. 2010. The effect of thymosin treatment of venous ulcers. *Ann. N. Y. Acad. Sci.* **1194:** 207–212.
39. Falanga, V., D. Margolis & O. Alvarez. 1998. Rapid healing of venous ulcers and lack of clinical rejection with an allogeneic cultured human skin equivalent. *Arch. Dermatol.* **134:** 293–300.
40. Perrin, M. & A.A. Ramelet. 2011. Pharmacological treatment of primary chronic venous disease: rationale, results and unanswered questions. *Eur. J. Vasc. Endovasc. Surg.* **41:** 117–125.
41. Roy, S., D. Patel & S. Khanna. 2007. Transcriptome-wide analysis of blood vessels laser captured from human skin and chronic wound-edge tissue. *Proc. Natl. Acad. Sci. USA* **104:** 14472–14477.
42. Peirce, S.M., T.C. Skalak & G.T. Rodeheaver. Ischemia–reperfusion injury in chronic pressure ulcer formation: a skin model in the rat. *Wound Repair Regen.* **8:** 68–76.
43. Reid, R.R., A.C. Sull & J.E. Mogford. 2004. A novel murine model of cyclical cutaneous ischemia–reperfusion injury. *J. Surg. Res.* **116:** 172–80.
44. Stadler, I., R.Y. Zhang & P. Oskoui. 2004. Development of a simple, noninvasive, clinically relevant model of pressure ulcers in the mouse. *J. Invest. Surg.* **17:** 221–227.
45. Lowthian, P. 1997. Notes on the pathogenesis of serious pressure sores. *Br. J. Nurs.* **6:** 07–12.
46. Salcido, R., A. Popescu & C. Ahn. Animal models in pressure ulcer research. *J. Spinal Cord Med.* **30:** 107–116.
47. Bergan, J.J., L. Pascarella & G.W. Schmid-Schönbein. 2008. Pathogenesis of primary chronic venous disease: insights from animal models of venoushypertension. *J. Vasc. Surg.* **47:** 183–192.
48. Takase, S., L. Pascarella & J.J. Bergan. 2004. Hypertension-induced venous valve remodeling. *J. Vasc. Surg.* **39:** 1329–1334.
49. Takase, S., L. Pascarella & G.W. Schmid-Schönbein. 2005. Venous hypertension and the inflammatory cascade: major manifestations and trigger mechanisms. *Angiology* **56:** S3–S10.
50. Lalka, S.G., J.L. Unthank & J.C. Nixon. 1998. Elevated cutaneous leukocyte concentration in a rodent model of acute venous hypertension. *J. Surg. Res.* **74:** 59–63.
51. Perez, R. & S.C. Davis. 2008. Relevance of animal models for wound healing. *Wounds* **20:** 3–8.
52. Singer, A.J. & S.A. McClain. 2003. Development of a porcine excisional wound model. *Acad. Emerg. Med.* **10:** 1029–1033.
53. Sullivan, T.P., W.H. Eaglstein & S.C. Davis. 2001. The pig as a model for human wound healing. *Wound Repair Regen.* **9:** 66–76.
54. Wang, J.F., M.E. Olson & C.R. Reno. 2001. The pig as a model for excisional skin wound healing: characterization of the molecular and cellular biology, and bacteriology of the healing process. *Comp. Med.* **51:** 341–348.
55. Heinrich, W., P.M. Lange & T. Stirtz. 1971. Isolation and characterization of the large cyanogen bromide peptides from the alpha1- and alpha 2- chains of pig skin collagen. *FEBS Lett.* **16:** 63–67.
56. Davidson, J.M. 1998. Animal models for wound repair. *Arch. Dermatol. Res.* **290:** S1–S11.

57. Davis, S.C., C. Ricotti & A. Cazzaniga. 2008. Microscopic and physiologic evidence for biofilm-associated wound colonization in vivo. *Wound Repair Regen.* **16:** 23–29.
58. Wassermann, E., M. van Griensven & K. Gstaltner. 2009. Chronic pressure ulcer model in the nude mouse. *Wound Repair Regen.* **17:** 480–484.
59. Roy, S., S. Biswas & S. Khanna. 2009. Characterization of a preclinical model of chronic ischemic wound. *Physiol. Genomics* **37:** 211–224.
60. Lin, F., A. Pandya & A. Cichowski. 2010. Deep tissue injury rat model for pressure ulcer research on spinal cord injury. *J. Tissue Viability* **19:** 67–76.

Ann. N.Y. Acad. Sci. ISSN 0077-8923

ANNALS OF THE NEW YORK ACADEMY OF SCIENCES
Issue: *Thymosins in Health and Disease*

The regenerative peptide thymosin β4 accelerates the rate of dermal healing in preclinical animal models and in patients

Terry Treadwell,[1] Hynda K. Kleinman,[2] David Crockford,[3] Mark A. Hardy,[4] Giorgio T. Guarnera,[5] and Allan L. Goldstein[2]

[1]Institute for Advanced Wound Care, Montgomery, Alabama. [2]Department of Biochemistry and Molecular Biology, The George Washington University School of Medicine and Health Sciences, Washington, DC. [3]RegeneRx Biopharmaceuticals Inc., Rockville, Maryland. [4]Department of Surgery, Columbia University, College of Physicians and Surgeons, New York, New York. [5]Istituto Dermatopadico dell'Immacolata, Rome, Italy

Address for correspondence: Hynda K. Kleinman, NIH, NIDCR, Building 30, Room 402, 30 Convent Dr MSC-4370, Bethesda, MD 20892-4370. hkleinman@dir.nidcr.nih.gov

Chronic nonhealing cutaneous wounds are a worldwide problem with no agent able to promote healing. A naturally occurring, endogenous repair molecule, thymosin beta 4 (Tβ4), has many biological activities that promote dermal repair. It is released by platelets at the site of injury and initiates the repair cascade. Tβ4 accelerated dermal healing of full-thickness punch wounds in various animal models, including normal rats and mice, steroid-treated rats, diabetic mice, and aged mice. Furthermore, in two phase 2 clinical trials of stasis and pressure ulcers, it was found to accelerate healing by almost a month in those patients that did heal. Tβ4 likely acts to repair and regenerate wounds by promoting cell migration and stem cell mobilization and differentiation, and by inhibiting inflammation, apoptosis, and infection. We conclude that Tβ4 is a multifunctional regenerative peptide important in dermal repair.

Keywords: dermal wounds; stasis ulcers; pressure ulcers; clinical trial; thymosin β4

Tβ4 activities in wounds

Chronic nonhealing cutaneous wounds are associated with significant patient morbidity and high healthcare costs.[1] Previously reported nonclinical studies suggest that Tβ4 may have the biological properties found useful clinically in these patient populations. Tβ4 is a highly conserved naturally occurring, water-soluble peptide.[2,3] In addition to being found in a majority of tissues and cell types, it is also found in the blood, and in other body fluids, including wound fluids, tears, saliva, and cerebrospinal fluid.[4–6] When platelets aggregate, they release both Tβ4 and transglutaminase (factor XIIIa) which crosslinks Tβ4 with the αC domains of fibrin.[7] Tβ4, now bound covalently to the clot matrix, is chemotactic for the directional migration of endothelial cells and keratinocytes, which migrate rapidly into the wound site, allowing the healing process to begin.[7] Tβ4 is increased significantly in wound fluids following surgery[8] and during ischemic injury to the heart.[9]

In animal studies, Tβ4 has multiple, well-defined biological activities that contribute significantly to reduce tissue damage following injury and to promote tissue repair.[10] Such activities include G-actin sequestration, promotion of cell and tissue survival, inhibition of inflammatory cell infiltration, down-regulation of cytokines and chemokines, enhancement of stem cell recruitment and differentiation, maturation of blood vessels, and prevention of scar formation.[10–17] These activities have been observed in animal studies following dermal, ocular, cardiovascular, and CNS injuries.[11–13,15–21] In particular, Tβ4 has accelerated ocular repair in both preclinical models and in two phase 2 and compassionate

doi: 10.1111/j.1749-6632.2012.06717.x

use trials with patients with dry eye and with neurotrophic keratitis, respectively. In the eye, it accelerates corneal repair in both alkali and heptanol debridement injuries in rodent models by decreasing inflammation and by increasing epithelial cell migration.[12,20] Furthermore, in early clinical trials, an ophthalmic eye-drop formulation of Tβ4 has shown promise in promoting ocular repair in patients with neurotropic keratitis and nonhealing corneal wounds.[22,23] In preclinical models of heart attack or reperfusion injury, Tβ4 reduced scar volume and improved heart function and recovery in both mice and pigs. In the heart, Tβ4 promotes repair after coronary artery ligation in rodents and after reperfusion injury in pigs by reducing inflammation and fibrosis and by increasing stem cell recruitment and differentiation.[21,24] In preclinical CNS models of traumatic brain injury, Tβ4 treatment has resulted in stem cell activation and differentiation, decreased inflammation, and functional recovery.[11,18] These data support the repair and regenerative properties of Tβ4 in multiple tissues with different types of injuries.[24]

Tβ4 in dermal injuries in preclinical models

Dermal repair with Tβ4 has been extensively studied in various preclinical rodent models with full-thickness punch wounds. In rodent models, topical treatment of Tβ4 promotes full-thickness dermal healing in normal rats and mice as well as in diabetic mice, aged mice, and steroid-treated rats.[10,17,19] Tβ4 has multiple activities that promote dermal repair (Table 1). Tβ4 acts by increasing keratinocyte migration (reepithelialization), organizing the distribution of collagen (regranulation), and stimulating angiogenesis (new blood vessel formation). It also reduces inflammation and has antimicrobial activity. Most importantly, the collagen fibers are better organized due to a reduction in the presence of myofibroblasts.[16] These scar-forming cells as defined by alpha smooth muscle actin staining are absent from Tβ4-treated wounds. Laminin-5 production is also increased by Tβ4. Laminin-5 is important in maintaining cell to cell and cell to matrix contacts.[25] These data on the positive effects of Tβ4 on dermal healing combined with the beneficial effects in other tissues discussed earlier provide a sound rationale for using Tβ4 to treat patients with chronic nonhealing dermal injuries, such as stasis and pressure ulcers.

Table 1. Biological activities of Tβ4 important in dermal healing

- Cell and keratinocyte migration
- Blood vessel formation (angiogenesis)
- Anti-inflammatory
- Antimicrobial
- Increased laminin-5 synthesis
- Increased rate of collagen deposition and organization
- Decreased myofibroblasts

Tβ4 in dermal injuries: clinical trials

In phase 1 clinical trials Tβ4 was deemed safe and well tolerated, when applied topically to the skin or administered intravenously in healthy human volunteers.[24,26] No adverse effects were noted in these studies and also in animal toxicity tests and in human phase I studies with intravenously injected Tβ4.[26] For these studies, very high doses beyond that to be used on patients were used confirming its safety. These safety studies demonstrate that Tβ4 is a safe, naturally occurring molecule for human use.

Two independent randomized, double-blind, placebo-controlled, dose-response phase 2 clinical trials evaluated the safety and efficacy of RGN-137 (the topical gel formulation of Tβ4) in the treatment of 143 total patients with chronic cutaneous (stage III/IV) pressure ulcers (full thickness) and venous stasis ulcers in which the majority of the patients had varicose veins and an open ulceration.[24,27] Among those patients with wounds that healed during the 84-day treatment period, Tβ4 at the mid-dose increased the rate of complete wound healing in both patient populations. The rate of healing in the mid-dose Tβ4-treated patients was approximately one month faster than either the placebo- or other RGN-137 dose-treated wounds.

Stage III and stage IV phase 2 pressure ulcer trial

Patients were divided into four groups and were treated with either placebo, 0.01%, 0.02%, or 0.1% Tβ4 gel. The percentage of patients who healed completely at day 84 was similar in the placebo and in the 0.02% Tβ4 treatment group at 17% while the lower and higher RGN-137 treatment doses had fewer patients who healed completely. The reason for this

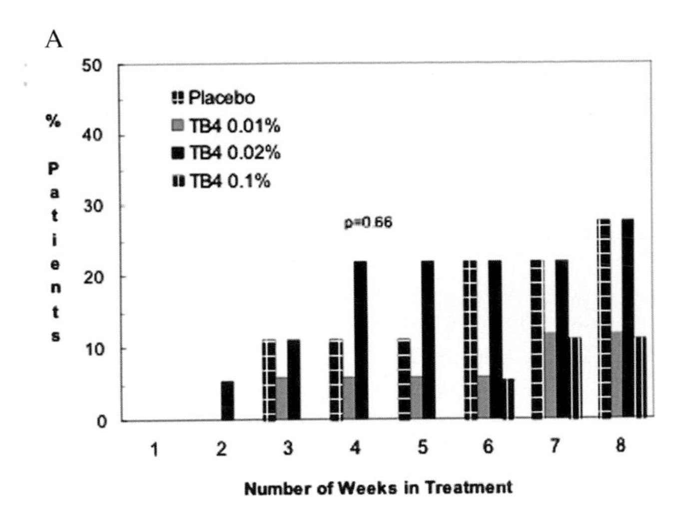

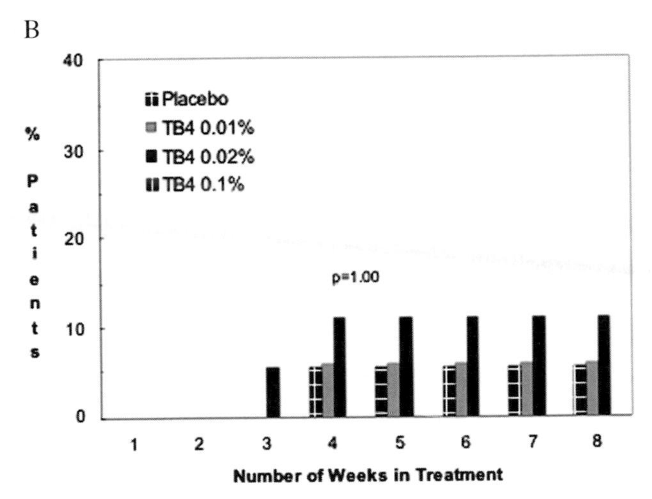

Figure 1. (A) Incidence with time of pressure ulcer wounds achieving 80% closure. (B) Incidence with time of pressure ulcer wounds achieving 100% closure. *P* value was based on a 2-sided Fisher's exact test versus placebo.

lower rate of healing in these two bracketing doses is unclear but was observed in the venous stasis ulcer trial as well and could be related to the bell curve of activity associated with the many receptor-mediated biological responses.[28] An 80% wound closure from baseline healing was observed as early as two weeks in the 0.02% treatment group compared to placebo. This healing continued at a faster rate until week six at which time some wounds that had been present as long as two years, healed completely. The healing of the placebo eventually caught up with the 0.02% RGN-137 treatment during the remainder of the 84-day treatment (Fig. 1A). Likewise, at 100% closure, the 0.02% Tβ4 dose showed accelerated healing over the placebo as early as three weeks (Fig. 1B). The time to healing among the wounds that did heal was much shorter in the 0.02% treatment group over the placebo group, with a median healing time of 22 days versus 57 days for the placebo (Table 2). The mean was 36 ± 25 days for the 0.02% RGN-137–treated group versus 51 ± 24 days for the placebo. The highest dose group did not have any patients who healed. The lowest dose group showed healing similar to those treated with the mid-dose, 0.02% RGN-137.

When the pressure ulcer wounds were analyzed by stage, more rapid healing was observed in the stage III group treated with 0.02% RGN-137. Healing in stage IV patients occurred only in the 0.02% RGN-137 group. These data show that the mid-dose (0.02%) of RGN-137 accelerates pressure ulcer healing by more than a month.

Table 2. Time (days) to healing during 84-day treatment

Pressure ulcer	Parameter	Placebo	0.01%	0.02%	0.1%
N (healed/total)	*N*	3/18	1/17	3/18	0/18
Time (days) to healing among all wounds healed	Median	57	22	22	None healed
	P value		0.37	0.38	
Time (days) to healing among stage III wounds healed	Median	57	22	22	None healed
Time (days) to healing among stage IV wounds healed	Median	None healed	None healed	64	None healed
Venous stasis ulcer	**Parameter**	**Placebo**	**0.01%**	**0.03%**	**0.1%**
N (healed/total)	*N*	4/17	3/19	6/18	3/18
Time (days) to healing among all wounds healed	Median	71	85	39	77
	P value		0.38	0.20	1.00
Time (days) to healing among ≤ 3 cm² wounds healed	Median	21	49	28	NA
Time (day) to healing among >3 cm² wounds healed	Median	78	88	49*	77
Pooled data	**Parameter**	**Placebo**	**0.01%**	**0.02–0.03%**	**0.1%**
N (healed/total)	*N*	7/35	4/36	9/36	3/36
Baseline wound area (cm²) of all wounds healed	Median	7.00	5.55	4.20	4.50
Time (days) to healing among all wounds healed	Median	63	67	28	77
	P value		0.64	0.14	0.73

NOTE: Wilcoxon rank-sum test versus placebo (2-sided).
*$P < 0.05$ (2-sided) against placebo (median test).
NA, not applicable.

Phase 2 venous stasis ulcer trial

Patients were divided into four groups and received the placebo, 0.01%, 0.03%, or 0.1% Tβ4 in the gel formulation. The number of patients who healed was 33% in the 0.03% RGN-137 treatment group versus 24% in the placebo. Again, the low and high doses of RGN-137 groups had reduced healing relative to the placebo and the 0.03% RGN-137 treatment groups (Table 2). Healing was observed to be greatest in patients treated with the RGN-137 at the 0.03% treatment dose over both the placebo and the other doses, and this was seen as early as one week after initiation of treatment when examined for 80% or 100% complete wound closure (Fig. 2A and B). The healing continued to be accelerated for an additional nine weeks as compared to the placebo. The time to healing among the wounds that completely healed was much shorter in the 0.03% treatment group whose median baseline ulcer area was lower as compared to the placebo with the median healing time of 39 days versus 71 days for 0.03% Tβ4 versus the placebo (Table 2). The mean was 37 ± 20 days for the 0.03%-treated group versus 62 ± 28 days for the placebo. The rates of wound healing were further analyzed based on wound surface area at either <3 cm² or 3 cm². As shown in Table 2, the larger wounds had, as expected, a longer healing time, whereas the smaller wounds had a faster healing time, as compared to the observed time to healing for all the wounds. Among the wounds that healed in patients whose wound size was greater than 3 cm², the median time to healing with 0.03% Tβ4 was

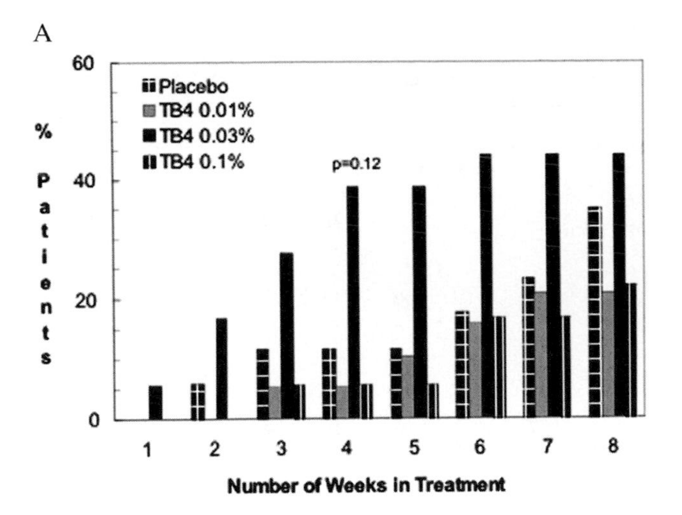

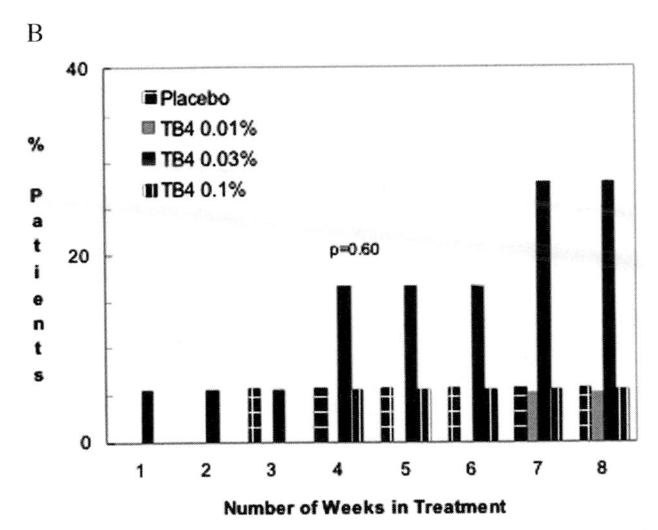

Figure 2. (A) Incidence with time of venous stasis ulcers achieving 80% closure. (B) Incidence with time of venous stasis ulcers achieving 100% closure. *P* value was based on a 2-sided Fisher's exact test versus placebo.

significantly faster than those treated with the placebo: 49 days versus 78 days ($P < 0.05$). These data, which show acceleration of healing of venous stasis ulcers by approximately one month, along with those described earlier for healing of pressure ulcers support the superior effectiveness of the mid-dose of RGN-137 (0.03% Tβ4) for the venous stasis ulcers.

Pooled data analysis (pressure ulcer and venous stasis ulcer)

The pooled data from both phase 2 trials was analyzed to determine if the effectiveness of each of the two mid-doses held across both trials. Despite the fact that the mid-doses were not identical (0.02 and 0.03% RGN-137), they were combined for the purpose of this analysis. The healing rate in patients in both studies was increased at the 80% and 100% closure measurements for the mid-dose (Fig. 3A and B) at all time points shown up to eight weeks of treatment. The combined median time to healing was 28 days for the mid-dose and 63 days for the placebo (Table 2). The combined mean time to healing for the mid dose was 37 ± 20 days versus 57 ± 25 days for the placebo. These data demonstrate that RGN-137 was effective at the mid-dose of 0.02–0.03% in accelerating the rate of wound healing of two very different types of cutaneous wounds that have different

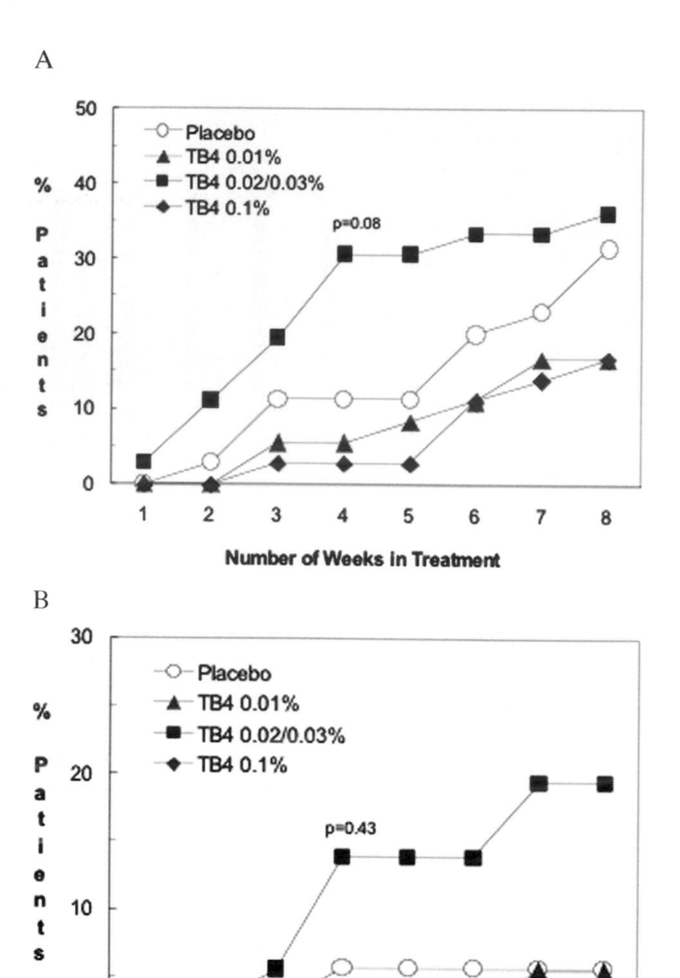

Figure 3. (A) Incidence with time of pooled data from pressure ulcers and venous stasis ulcers achieving 80% closure. (B) Incidence with time of pooled data from pressure ulcers and venous stasis ulcers achieving 100% closure. *P* value was based on a 2-sided Fisher's exact test versus placebo.

etiologies. The mid-dose (0.02/0.03%) showed considerable acceleration of healing relative to the other doses (0.01 and 0.10%) and to the placebo in both trials. It is not clear why the lower and higher doses were less effective/ineffective but Tβ4 is a biologically active protein with a defined receptor.[29] Such bioactive molecules can have a narrow effective dose range and differing responses seen at different concentration are common to many signaling molecules for example, the wound-healing molecule TGF-β3 demonstrates a bell-shaped dose response curve.[28] $T\beta_4$ also showed a similar bell-shaped curve in mice with multiple sclerosis (Morris, unpublished) and a phase 2 trial for dry eye patients (unpublished).

Summary and conclusions

Dermal healing is a complex process. Tβ4, a small naturally occurring wound repair peptide, is released early in the wound by the platelets, which arrive in the wound to clot and begin the healing process. This protein is multifunctional and has many activities important for wound repair and regeneration. Studies with various animal models of normal and impaired healing support its clinical potential. Furthermore, studies of injuries in the brain, heart, and eye treated with Tβ4 show consistent beneficial effects.

Two phase 2 clinical dry eye trials and compassionate use studies showed positive effects on eye recovery. Two preliminary phase 2 clinical trials with Tβ4 used to treat stasis and pressure ulcers also showed beneficial effects. Among those patients whose wounds healed, the accelerated healing resulted in a reduction of approximately one month of wound care needed to treat the wounds. Although less than half of the patients in the mid-dose treatment group healed, treatment with Tβ4 could be of significant medical, economic, and social benefit given the high pain, morbidity, and costs associated with treatment of these chronic wounds, which are generally found in elderly patients. At this time, it is not clear which patients would heal faster with Tβ4 and which are nonresponders or why this occurs. Future studies would require treatment of all patients until some way could be determined to identify the healers or the nonhealers. In the United States, it has been estimated that 3–10% of hospitalized patients have pressure ulcers and this number increases significantly (20–32%) in elderly patients who are hospitalized with long-term disabilities. In 2006, for example, the costs of hospitalization related to pressure ulcers in the United States alone was estimated to be over 11 billion dollars and affects more than 1% of the population.[30] Reduction by one month in the time necessary for treatment of each of the chronic wounds studied in these phase 2 trials would provide a major advance in the care of patients. These patients present with such wounds that are generally difficult to heal and particularly in the case of pressure ulcers, which do not heal for many years, may become infected, and often lead to death, due to septicemia.

Conflicts of interest

David Crockford holds stock and is employed by RegeneRx Biopharmaceuticals Inc. (RGN), which is developing Tβ4 for therapeutic uses. Allan Goldstein holds stock and is the Chief Scientific Officer for RGN. Hynda Kleinman is a consultant for RGN and has patent rights to the compound for tissue repair.

References

1. Baranaski, S. 2006. Raising awareness of pressure ulcer prevention and treatment. *Adv. Skin Wound Care* **19:** 398–400.
2. Low, T.L.K., S.K. Hu & A.L. Goldstein. 1981. Complete amino acid sequence of bovine thymosin β_4: a thymic hormone that induces terminal deoxynucletidyltransferase activity in thymocyte populations. *Proc. Natl. Acad. Sci.* **78:** 1166–1170.
3. Wang, S.S., B.S.H. Wang, J.K. Chang, *et al.* 1981. Synthesis of thymosin β_4. *Int. J. Peptide Protein Res.* **18:** 413–415.
4. Frohm, M., H. Gunn, A.C. Bergman, *et al.* 1996. Biochemical and antibacterial analysis of human wound and blister fluid. *Eur. J. Biochem.* **237:** 86–92.
5. Badamchian, M., A.A. Damavandy, H. Damavandy, *et al.* 2007. Identification and quantification of thymosin beta 4 in human saliva and tears. *Ann. N. Y. Acad. Sci.* **1112:** 458–465.
6. Urso, E., M. Le Pera, S. Bossio, *et al.* 2010. Quantification of thysmosinbeta(4) in human cerebrospinal fluid using matrix-assisted laser desorption/ionization time-of-flight mass spectrometry. *Anal. Biochem.* **402:** 13–19.
7. Huff, T., A.M. Otto, C.S. Muller, *et al.* 2002. Thymosin beta4 is released from human blood platelets and attached by factor XIIIa (transglutaminase) to fibrin and collagen. *FASEB J.* **16:** 691–696.
8. Bodendorf, S., G. Born & E. Hannappel. 2007. Determination of thymosin β_4 and protein in human wound fluid after abdominal surgery. *Ann. N. Y. Acad. Sci.* **1112:** 418–424.
9. Shah, A.P., S.T. Youngquist, C.D. McClung, *et al.* 2011. Markers of progenitor cell recruitment and differentiation rise early during ischemia and continue during resuscitation in a porcine acute ischemia model. *J. Interferon Cytokine Res.* **31:** 679–684.
10. Goldstein, A.L., E. Hannappel & H.K. Kleinman. 2005. Thymosin beta 4: actin-sequestering protein moonlights to repair injured tissues. *Trends Mol. Med.* **11:** 421–429.
11. Morris, D.C., M. Chopp, L. Zhang, *et al.* 2010. Thymosin beta 4 improves functional neurological outcome in a rat model of embolic stroke. *Neuroscience* **169:** 674–682.
12. Sosne, G., F.A. Szliter, R. Barrett, *et al.* 2002. Thymosin beta 4 promotes corneal wound healing and decreases inflammation in vivo following alkali injury. *Exp. Eye Res.* **74:** 293–299.
13. Smart, N., C.A. Riseboro, A.A. Melville, *et al.* 2007. Thymosin beta 4 induces adult cardiac progenitor mobilization and neovascularization. *Nature* **445:** 177–182.
14. Grant, D.S., J.L. Kinsella, M.C. Kibbey, *et al.* 1995. A novel role for thymosin beta 4: a Matrigel-induced gene involved in endothelial cell differentiation and angiogenesis. *J. Cell Science* **108:** 3685–3694.
15. Bock-Marquette, I., A. Saxena, M.D. White, *et al.* 2004. Thymosin beta 4 activates integrin-linked kinase and promotes cardiac cell migration, survival and cardiac repair. *Nature* **432:** 466–472.
16. Ehrlich, H.P. & S.W. Hazard 3rd. 2007. Thymosin beta4 enhances repair by organizing connective tissue and preventing the appearance of myofibroblasts. *Ann. N. Y. Acad. Sci.* **1194:** 118–124.

17. Philp, D., M. Badamchian, B. Scheremeta, *et al.* 2003. Thymosin beta 4 and a synthetic peptide containing its actin-binding domain promote dermal wound repair in db/db diabetic mice and in aged mice. *Wound Rep. Reg.* **11:** 19–24.
18. Zhang, J., Z.G. Zhang, D. Morris, *et al.* 2009. Neurological functional recovery after thymosin beta 4 treatment in mice with experimental auto encephalomyelitis. *Neuroscience* **164:** 1887–1893.
19. Li, X., L. Zheng, F. Peng, *et al.* 2007. Recombinant thymosin beta 4 can promotoe full thick cutaneous wound healing. *Protein Expr.Purif.* **56:** 229–236.
20. Sosne, G., A. Siddiqi & M. Kurpakus-Wheater. 2004. Thymosin-beta4 inhibits corneal epithelial cell apoptosis after ethanol exposure in vitro. *Invest. Ophthalmol. Vis. Sci.* **45:** 1095–100.
21. Hinkel, R., C. El-Aouni, T. Olson, *et al.* 2008. Thymosin beta4 is an essential paracrine factor of embryonic endothelial progenitor cell-mediated cardioprotection. *Circulation* **117:** 2232–2240.
22. Sosne, G., P. Qiu, M. Kurpakus-Wheater & H. Matthew. 2010. Thymosin beta4 and corneal wound healing: visions of the future. *Ann. N. Y. Acad. Sci.* **1194:** 190–198.
23. Dunn, S.P., D.G. Heideman, C.Y. Chow, *et al.* 2010. Treatment of chronic non-healing neurotrophic corneal epithelial defects with thymosin beta 4. *Arch. Opthalmol.* **128:** 636–638.
24. Crockford, D., N. Turjman, C. Allan & J. Angel. 2010. Thymosin beta4: structure, function, and biological properties supporting current and future clinical applications. *Ann. N. Y. Acad. Sci.* **1194:** 179–189.
25. Sosne, G., L. Xu, L. Prach, *et al.* 2004. Thymosin beta 4 stimulates laminin-5 synthesis indepdent of TGF-beta. *Exp. Cell Res.* **293:** 175–183.
26. Ruff, D., D. Crockford, G. Girardi & Y. Zhang. 2010. A randomized, placebo-controlled, single and multiple dose study of intravenous thymosin β_4 in healthy volunteers. *Ann. N. Y. Acad. Sci.* **1194:** 223–229.
27. Guarnera, G., A. DeRosa & R. Camerini. 2010. The effect of thymosin treatment of venous ulcers. *Ann. N. Y. Acad. Sci.* **1194:** 207–212.
28. Occleston, N.L., S.O. O'Kane, H.G. Laverty, *et al.* 2011. Discovery and development of avoterim (recombinant human transforming growth factor beta 3): a new class of prophylactic therapeutic for the improvement of scarring. *Wound Rep. Regen.* **19:** S38–S48.
29. Freeman, K.W., B.R. Bowman & B.R. Zetter. 2010. Regenerative protein thymosin beta-4 is a novel regulator of purinergic signaling. *FASEB J.* **25:** 907–915.
30. Russo, C.A., C. Steiner & W. Spector. 2008. Hospitalizations related to pressure ulcers, 2006. In *HCUP Statistical Brief #64.* Agency for Healthcare Research and Quality. Rockville. MD. Available at: http://www.hcup-us.ahrg.gov/reports/statbriefs/sh64.pdf.

Ann. N.Y. Acad. Sci. ISSN 0077-8923

ANNALS OF THE NEW YORK ACADEMY OF SCIENCES
Issue: *Thymosins in Health and Disease*

Thymosin β4: a potential novel dry eye therapy

Gabriel Sosne,[1] Ping Qiu,[1] George W. Ousler 3rd,[2] Steven P. Dunn,[3] and David Crockford[4]

[1]Department of Ophthalmology, Kresge Eye Institute, Wayne State University School of Medicine, Detroit, Michigan. [2]Ora, Andover, Massachusetts. [3]Michigan Cornea Consultants, Southfield, Michigan. [4]RegeneRx Biopharmaceuticals Inc., Rockville, Maryland

Address for correspondence: Gabriel Sosne, M.D., Kresge Eye Institute, Wayne State University School of Medicine, 4717 St. Antoine, Detroit, MI 48201. gsosne@med.wayne.edu

The purpose of this manuscript is to review the clinical entity of dry eye syndrome (DES) and to provide a scientific basis and rationale for the usage of thymosin beta 4 (Tβ4) as a novel therapy for DES. DES is a common disorder affecting an estimated 25–30 million people in the United States alone and is characterized by inflammation of the ocular surface. Consequently, patients can suffer from burning, irritation, severe discomfort, foreign body sensation, and blurry and decreased vision. Recent animal studies of DES demonstrate that Tβ4 eye drops significantly reduce corneal fluorescein staining, indicating improved wound healing. Based on previous studies, there is clear support for further clinical investigation and development of Tβ4 as a novel, safe, and effective agent to treat dry eye. Herein, we discuss the scientific and clinical rationales that make Tβ4 a potential ideal candidate therapeutic for DES.

Keywords: thymosin β4; inflammation; dry eye disease; ocular surface; cornea

Dry eye disease: overview and pathophysiology

Dry eye syndrome (DES) is one of the most common conditions seen by ophthalmologists and affects 15% of those from 65–84 years of age, or 4.3 million Americans.[1] As DES is one of the most common ophthalmic medical problems, complaints of dry eye are among the most common reasons patients seek help from eye doctors. Many patients suffering from dry eye live with constant pain and irritation, and the condition can be quite debilitating. Recent studies suggest that the impact on quality of life from dry eye disease is approximately equal to that of angina.[2] Until recently, DES was related dysfunction of the lacrimal gland and treatments were aimed at lubricating and hydrating the ocular surface. However, dry eye disease is now considered a complex inflammatory syndrome that involves a number of dysregulated biochemical, neurological, and hormonal pathways. We now understand that the tear film is secreted reflexively from the lacrimal functional unit, which is composed of the ocular surface tissues, the lacrimal glands, and their interconnecting sensory and autonomic innervation.[3] Dysfunction of this integrated functional unit may result in a decrease in the quantity of tears and can lead to changes in tear composition that result in loss of tear film integrity and promote inflammation. The pathophysiology of dry eye includes immune-mediated inflammation involving both the ocular surface and the lacrimal gland.

Elevated tear film osmolarity has been identified as a common feature of dry eye disease.[4,5] Significantly increased tear osmolarity has been reported in a variety of dry eye and ocular surface diseases, including lacrimal gland secretory dysfunction, meibomian gland disease, contact lens wear, exposure keratopathy, and neurotrophic keratopathy.[5–9] Hyperosmolar stress activates key inflammatory pathways in epithelial and inflammatory cells, including the transcriptional factor nuclear factor kappa B (NF-κB) and IL-8.[10] NF-κB signaling plays a central role in generating a host of proinflammatory cytokine, chemokine, and adhesion factor responses that mediate tissue responses in inflammation. Consistent with this, TNF-α causes translocation of

doi: 10.1111/j.1749-6632.2012.06682.x

Table 1. Key $T\beta_4$ activities important in wound repair in the eye

Activity	Significance/importance
Anti-inflammation	Prevents swelling, pain, tissue damage
Promotes cell migration and tissue integrity	Promotes reepithelialization and stem cell recruitment (presumptive) and tight cell contacts
Protects from burns and cytotoxic agents	Prevents alkali, alcohol burns, and toxicity because of benzalkonium, glutamate, and kainic acid
Antiapoptosis/survival	Prevents cell death from infection, toxins, and increased cell survival after trauma
Antifibrotic	Prevents scarring and tissue damage, reverses fibrosis, and maintains tissue function

NF-κB to the nucleus where it can actively modulate gene expression.[11]

Although knowledge of the pathology of dry eye disease has improved significantly during recent years, the mainstay of treatment, ocular lubrication, provides only palliative relief at best to patients with severe dry eye disease. Therefore, attention has turned to immunomodulation as a therapeutic approach for severe dry eye disease, especially for that seen in graft versus host disease (GVHD), as well as autoimmune diseases like Sjögren's syndrome, rheumatoid arthritis, and systemic lupus erythematosus (SLE). Sjögren's syndrome is a chronic autoimmune disease characterized by focal lymphocytic infiltration and destruction of salivary and lacrimal gland tissue.[12] The prevalence of Sjögren's syndrome may approach that of rheumatoid arthritis, which affects between 1% and 3% of the general population and occurs almost exclusively in women (>90%).[13,14] In addition to primary Sjögren's syndrome, secondary Sjögren's syndrome occurs in association with other autoimmune diseases such as GVHD, rheumatoid arthritis, and SLE. Sjögren's syndrome is an important cause of aqueous deficient dry eye known as keratoconjunctivitis sicca (KCS).[15,16] The precise etiology and pathogenic mechanisms of Sjögren's syndrome are poorly understood at present and to date, there is no cure.

Histopathologic evaluation of exocrine gland tissue in Sjögren's syndrome shows predominant $CD4^+$ T lymphocytic infiltration and hyperreactive autoreactive B lymphocytes. Immune-mediated dysfunction and destruction of the glandular epithelial cells are responsible for the decline in tear and saliva production.[17,18] Additional studies have found that immune-mediated inflammation plays a role in the pathogenesis of dry eye, not only in Sjögren's syndrome, but also in postmenopausal and age-related dry eye. For example, conjunctival T lymphocytic infiltration and upregulation of epithelial cell MHC Class II and adhesion molecules have been demonstrated in both non-Sjögren's and Sjögren's syndrome KCS.[19–21] Conjunctival squamous metaplasia, a consequence of longstanding disease, is characterized by loss of mucin-producing goblet cells, increased epithelial cell cytoplasmic–nuclear ratio, and keratinization.[22]

Increased levels of inflammatory cytokines have also been demonstrated, especially IL-1, IL-6, and TNF-α in the lacrimal glands, conjunctival epithelium, and/or tear fluid of dry eye patients with or without Sjögren's syndrome.[21,23–25] Recent evidence suggests that elevated levels of IL-1β may impair the secretory function of the lacrimal gland and inhibit neurally mediated lacrimal gland secretion.[26] Other evidence includes increased expression of immune activation molecules such as HLA-DR, ICAM-1, CD-40, and CD-40 ligand on the ocular surface of patients with dry eye and it was demonstrated that the balance of cytokines in the tear fluid and conjunctival epithelium is altered in Sjögren's syndrome KCS.[22] Reports indicate that Sjögren's syndrome patients showed decreased tear fluid epidermal growth factor and significantly increased levels of IL-1α, IL-6, IL-8, TNF-α, and TGF-β1 in their conjunctival epithelium compared to control patients.[22] Animal models of autoimmune lacrimal gland disease have been used to explore the role of proinflammatory cytokines in the pathogenesis of KCS.[27] An understanding of the disease process of DES predicts that topical anti-inflammatory and immunomodulatory agents may be capable of normalizing the inflamed ocular surface and lacrimal glands.[28]

Pharmacologic therapies for dry eye

Presently, the most widely used therapy for dry eye is topical artificial tear replacement. According to the pharmaceutical company Allergan, "approximately 60 million people worldwide use artificial tears. It is estimated that over one million suffer from dry eye disease in the United States. Given the large number of individuals suffering from ocular surface disease, it is believed that the therapeutic segment of the dry eye market could grow to be as large as US$300– $500 million over the next three to five years" (www.restasis.com). By hydrating and lubricating the ocular surface with artificial tears, many patients report decreased ocular irritation and visual disturbance.[29,30] However, these therapies are only palliative, as they lubricate the surface without addressing the underlying disease process. Artificial tears have no direct anti-inflammatory effects, although they may secondarily decrease inflammation through their ability to lower tear osmolarity by diluting the concentration of inflammatory cytokines in the tear film.[31] Lubricant drops, gels, and ointments do not duplicate the complex composition of tears, which contain electrolytes, vitamins, growth factors, antiproteases, and a variety of antimicrobial proteins like lactoferrin and lysozyme. Furthermore, lubricant preparations are delivered intermittently, rather than continuously, as are natural tears, thus leaving the irritated ocular surface more prone to damage and infection. Therefore, comprehensive therapy for DES should be aimed at decreasing inflammation of the ocular surface and lacrimal glands, promoting healing of the epithelium, and normalizing tear composition and volume.

Available agents for treating severe DES, such as corticosteroids, have considerable side effects that render them therapeutically impotent in a number of patients. Corticosteroids are potent inhibitors of multiple inflammatory processes. They inhibit inflammatory cytokine and chemokine production, decrease the synthesis of matrix metalloproteinases (MMPs) and prostaglandins, decrease expression of cell adhesion molecules, and can stimulate lymphocyte apoptosis. Topical unpreserved methylprednisolone has been found useful in patients with DES from Sjögren's syndrome.[31] However, with chronic use, corticosteroids can have significant adverse effects, including elevated intraocular pressure, cataracts, impaired wound healing, potentiation of collagenases, and increased risk of infection, which limit their use in the therapy of dry eye.

Cyclosporin A (CSA) is a fungal-derived peptide that prevents activation and nuclear translocation of cytoplasmic transcription factors that are required for T cell activation and inflammatory cytokine production.[32] It is an established immunomodulator used to prevent transplant rejection, and as a treatment for autoimmune diseases. In an animal model of Sjögren's syndrome related dry eye, topically applied CSA was found to decrease the lymphocytic infiltration of lacrimal glands.[33] Topical CSA has been shown to increase tear production and reduce lacrimal gland inflammatory cell infiltrates in a KCS dog model.[34] In two phase III clinical trials, topical CSA has been shown to decrease corneal fluorescein staining, increase tear production (categorized Schirmer with anesthesia), and improve symptoms of dry eye.[35]

Topical 0.5% CSA (Restasis®), is approved for the treatment of moderate to severe dry eye. Fifteen percent of CSA-treated patients in the phase III clinical trials responded with a significant increase in tear production of 10 mm compared to 5% of vehicle-treated patients. However, 85% of CSA-treated patients did not experience this significant increase in tear production. Therefore, a therapeutic agent that could reliably decrease immune-mediated ocular surface damage without significant adverse effects in a large proportion of affected patients would constitute a major therapeutic advance, particularly for dry eye disease.

Tβ4 and dry eye: rationale

The need remains for an anti-inflammatory therapy that supports wound healing without complicating side effects for the treatment of diseases such as dry eye. Because Tβ4 has both anti-inflammatory and wound-healing properties, it is an ideal candidate for evaluation in an autoimmune model of dry eye. In brief, Tβ4 is a highly conserved 43 amino acid protein with wound-healing and immunomodulatory properties originally isolated from TF5 in 1981.[36] Until recently, Tβ4 was thought to function primarily as a G-actin–sequestering protein.[36–39] Tβ4 levels are highest in platelets and polymorphonuclear (PMN), which are among the

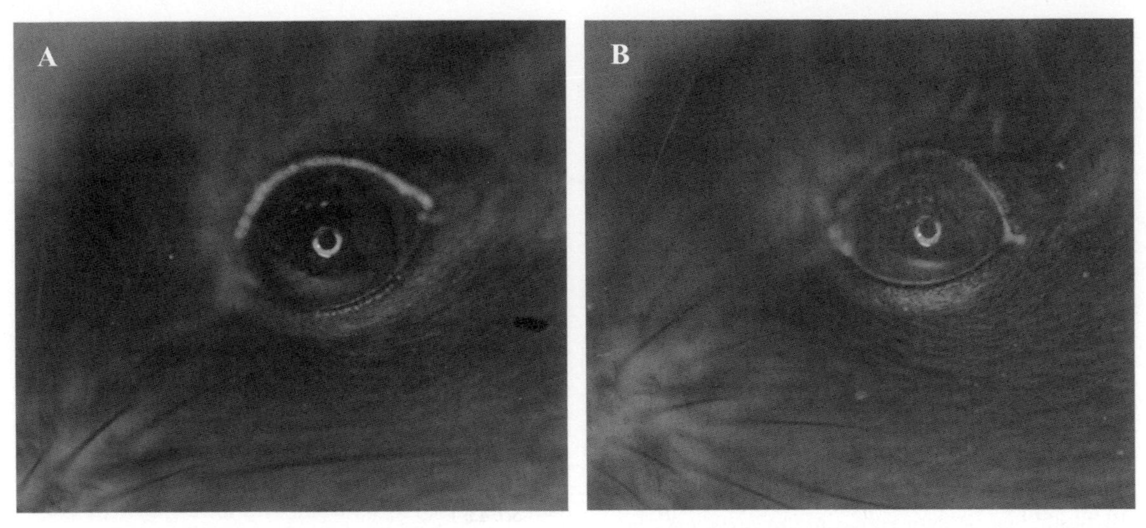

Figure 1. Topical eye drops (RGN-259) decrease corneal staining (indicating improved healing) in murine eyes exposed to the scopolamine and controlled adverse environment dry eye model (ORA, Andover, MA). Marked decrease in corneal fluorescein staining is demonstrated in Tβ4-treated (A) compared to control-treated eyes (B).

first formed elements/cells, respectively, to enter a wound and release their factors, some of which recruit additional cells to the wound site.[40] Previous studies reported that Tβ4 promotes full-thickness dermal wound repair in normal, steroid-treated, and diabetic rats.[41,42]

Previously, we demonstrated in corneal debridement and alkali injury models that Tβ4 accelerates corneal epithelial wound healing, reduces PMN corneal infiltration and decreases the expression of, for example, IL-1β, and IL-18, and chemokines, such as MIP-1α, MIP-1β, and MIP-2, as well as modulating MMP production (see Table 1).[43–45] Other recent studies on the immunomodulatory properties of Tβ suggest that Tβ4 lowers circulating levels of inflammatory cytokines and intermediates in a rat model of septic shock and may be useful in the therapy of septic shock.[46] Along with the observations that Tβ4 released from human blood platelets is a specific substrate for factor XIIIa and is cross-linked by factor XIIIa to fibrin clots and collagen, the literature suggests that Tβ4 may be useful in the treatment of both acute and chronic wounds.[47] Taken together, these findings provide a strong basis for our rationale to test the anti-inflammatory and wound-healing properties of Tβ4 on lacrimal gland and ocular surface tissue in autoimmune dry eye disease.

In addition, human patients suffering from neurotrophic corneal ulcers demonstrated marked healing after treatment with Tβ4 in a compassionate usage study without any adverse side effects reported.[48] In a recent exploratory study, researchers demonstrated in a murine model of dry eye disease using a controlled adverse environment that Tβ4 eye drops successfully decreased corneal fluorescein staining more than saline (negative) controls as well as Restasis and doxycycline (positive) controls (Fig. 1). These promising observations provide us with a solid rationale for using Tβ4 to promote ocular surface epithelial health and regulating inflammation in DES, especially when the dry eye is immune mediated.

Our overall hypothesis is that Tβ4 regulates corneal and conjunctival inflammation and thereby promotes ocular surface healing in dry eye disease. Because topical Tβ4 eye drops have no known adverse side effects in the eyes of animals and humans previously treated, we believe that Tβ4 should be considered as a novel therapy to treat DES. The effects of Tβ4 on ocular surface inflammation by examining the morphology of the ocular surface epithelium and the lacrimal gland in DES are poorly understood. We propose that Tβ4 alters production of cytokines, chemokines, and specific MMPs in DES, thereby decreasing ocular surface discomfort from inflammation and promoting wound healing.

Recent preclinical evaluations have demonstrated that Tβ4 promotes improved corneal epithelial intercellular adhesions following injury in animal models of dry eye.[49] These model studies show that Tβ4 reduced corneal staining (indicating improved healing) more than positive controls and demonstrated statistically significant reduction in staining compared to vehicle control. The results of these studies, in addition to data from compassionate-use studies in patients with nonhealing corneal surface defects, suggest that Tβ4 has a significant potential to be a novel safe and effective therapeutic treatment for dry eye. We believe that the potential of Tβ4 therapeutic effects on dry eye will stem foremost from its anti-inflammatory and wound-healing properties. As stated above, ocular surface inflammation plays a major role in corneal epithelial and sensory nerve disease of dry eye and causes the patient much pain and discomfort. Previously, we demonstrated that ectopic Tβ4 treatment suppresses TNF-α–mediated NF-κB activation and directly targets the NF- RelA/p65 subunit. We found that enforced expression of Tβ4 interferes with TNF-α–mediated NF-κB activation, as well as downstream IL-8 gene transcription, and that these activities are independent of the G-actin–binding properties of Tβ4. Tβ4 blocks RelA/p65 nuclear translocation and targeting to the cognate κB site in the proximal region of the IL-8 gene promoter after TNF-α stimulation.[50]

We propose that future studies elaborating the effects of Tβ4 treatment on lacrimal gland and ocular surface tissue changes, such as gland atrophy and fibrosis, conjunctival epithelial cell cytoplasmic–nuclear ratio, and keratinization (involucrin, filaggrin, keratin 1, and keratin 10), and the extent of inflammatory cellular (lymphocytic and PMN) infiltration, will provide crucial information as to the anti-inflammatory and pro-repair effects of Tβ4. In turn, these basic scientific findings will provide a strong basis for future mechanistic studies as well as translational clinical trials using Tβ4 as a novel treatment for DES.

Conflicts of interest

The authors declare no conflicts of interest.

References

1. Schein, O.D. *et al.* 1997. Prevalence of dry eye among the elderly. *Am. J. Ophthalmol.* **124:** 723–728.
2. Schiffman, R.M. *et al.* 2003. Utility assessment among patients with dry eye disease. *Ophthalmology* **110:** 1412–1419.
3. Pflugfelder, S.C. 2004. Antiinflammatory therapy for dry eye. *Am. J. Ophthalmol.* **137:** 337–342.
4. Farris R.L. 1994. Tear osmolarity–a new gold standard? *Adv. Exp. Med. Biol.* **350:** 495–503.
5. Gilbard, J.P. & R.L. Farris. 1983. Ocular surface drying and tear film osmolarity in thyroid eye disease. *Acta Ophthalmol.* **61:** 108–116.
6. Gilbard, J.P., S.R. Rossi & K.G. Heyda. 1989. Ophthalmic solutions, the ocular surface, and a unique therapeutic artificial tear formulation. *Am. J. Ophthalmol.* **107:** 348–355.
7. Gilbard, J.P. & K.R. Kenyon. 1988. Tear diluents in the treatment of keratoconjunctivitis sicca. *Ophthalmology* **92:** 646–650.
8. Gilbard, J.P. 1994. The tear film: pharmacological approaches and effects. *Adv. Exp. Med. Biol.* **350:** 377–384.
9. Gilbard, J.P. *et al.* 1989. Effect of punctal occlusion by Freeman silicone plug insertion on tear osmolarity in dry eye disorders. *CLAO J.* **15:** 216–218.
10. Nemeth, J. *et al.* 2002. High-speed videotopographic measurement of tear film build-up time. *Invest. Ophthalmol. Vis. Sci.* **43:** 1783–1790.
11. Tsubota, K. *et al.* 1999. Treatment of dry eye by autologous serum application in Sjögren's syndrome. *Br. J. Ophthalmol.* **83:** 390–395.
12. Strand, V. & N. Talal. 1979. Advances in the diagnosis and concept of Sjögren's syndrome (autoimmune exocrinopathy). *Bull. Rheum. Dis.* 30: 1046–1052.
13. Winer, S. *et al.* 2002. Primary Sjögren's syndrome and deficiency of ICA69. *Lancet* **360:** 1063–1069.
14. Azuma, T. *et al.* 2002. Identification of candidate genes for Sjogren's syndrome using MRL/lpr mouse model of Sjogren's syndrome and cDNA microarray analysis. *Immunol. Lett.* **81:** 171–176.
15. Fox, R.I. *et al.* 1999. Current issues in the diagnosis and treatment of Sjögren's syndrome. *Curr. Opin. Rheumatol.* **11:** 364–371.
16. Sullivan, D.A. 1997. Sex hormones and Sjögren's syndrome. *J. Rheumatol. Suppl.* **50:** 17–32.
17. Fox, R.I. *et al.* 1983. Characterization of the phenotype and function of lymphocytes infiltrating the salivary gland in patients with primary Sjogren syndrome. *Diagn. Immunol.* **1:** 233–239.
18. Adamson, T.D. *et al.* 1983. Immunohistologic analysis of lymphoid infiltrates in primary Sjogren's syndrome using monoclonal antibodies. *J. Immunol.* **130:** 203–208.
19. Andonopoulos, A.P. *et al.* 1989. Sjögren's syndrome in rheumatoid arthritis and progressive systemic sclerosis. A comparative study. *Clin. Exp. Rheumatol.* **7:** 203–205.
20. Jabs, D.A. *et al.* 1991. Murine models of Sjögren's syndrome: evolution of the lacrimal gland inflammatory lesions. *Invest. Ophthalmol. Vis. Sci.* **32:** 372–380.
21. Stern, M.E. *et al.* 2002. Conjunctival T-cell subpopulations in Sjögren's and non-Sjögren's patients with dry eye. *Invest. Ophthalmol. Vis. Sci.* **43:** 2609–2614.
22. Pflugfelder, S.C. *et al.* 1990. Conjunctival cytologic features of primary Sjögren's syndrome. *Ophthalmology* **97:** 985–991.
23. Bourcier, T. *et al.* 2000. Expression of CD40 and CD40 ligand in the human conjunctival epithelium. *Invest. Ophthalmol. Vis. Sci.* **41:** 120–126.
24. Brignole, F. *et al.* 2000. Flow cytometric analysis of inflammatory markers in conjunctival epithelial cells of

patients with dry eyes. *Invest. Ophthalmol. Vis. Sci.* **41:** 1356–1363.
25. Baudouin, C. *et al.* 1997. Flow cytometry in impression cytology specimens. A new method for evaluation of conjunctival inflammation. *Invest. Ophthalmol. Vis. Sci.* **38:** 1458–1464.
26. Zoukhri, D. & C.L. Kublin. 2002. Impaired neurotransmission in lacrimal and salivary glands of a murine model of Sjögren's syndrome. *Adv. Exp. Med. Biol.* **506 (Pt B):** 1023–1028.
27. Jabs, D.A. *et al.* 2001. Cytokines in autoimmune lacrimal gland disease in MRL/MpJ mice. *Invest. Ophthalmol. Vis. Sci.* **42:** 2567–2571.
28. Wilson, S.E. 2003. Inflammation: a unifying theory for the origin of dry eye syndrome. *Manag Care.* **12**(Suppl. 12): 14–19.
29. Nelson, J.D. & R.L. Farris. 1988. Sodium hyaluronate and polyvinyl alcohol artificial tear preparations. A comparison in patients with keratoconjunctivitis sicca. *Arch. Ophthalmol.* **106:** 484–487.
30. Toda, I., N. Shinozaki & K. Tsubota. 1996. Hydroxypropyl methylcellulose for the treatment of severe dry eye associated with Sjögren's syndrome. *Cornea* **15:** 120–128.
31. Pflugfelder, S.C. 2003. Anti-inflammatory therapy of dry eye. *Ocul. Surf.* **1:** 31–36.
32. Matsuda, S. 2002. Gamma knife radiosurgery for trigeminal neuralgia: the dry-eye complication. *J. Neurosurg.* **97**(Suppl. 5): 525–528.
33. Tsubota, K. 1998. Tear dynamics and dry eye. *Prog. Retin. Eye Res.* **17:** 565–596.
34. Stern, M.E. *et al.* 1998. The pathology of dry eye: the interaction between the ocular surface and lacrimal glands. *Cornea* **17:** 584–589.
35. Sall, K.N. *et al.* 2000. Two multicenter, randomized studies of the efficacy and safety of cyclosporine ophthalmic emulsion in moderate to severe dry eye disease. CsA Phase 3 Study Group. *Ophthalmology* **107:** 631–639.
36. Low, T.L. *et al.* 1981. Complete amino acid sequence of bovine thymosin beta 4: a thymic hormone that induces terminal deoxynucleotidyl transferase activity in thymocyte populations. *Proc. Natl. Acad. Sci. U S A* **78:** 1162–1166.
37. Low, T.L. & A.L. Goldstein. 1984. Thymosins: structure, function and therapeutic applications. *Thymus* **6:** 27–42.
38. Yu, F.X. *et al.* 1994. Effects of thymosin beta 4 and thymosin beta 10 on actin structures in living cells. *Cell Motil. Cytoskeleton* **27:** 13–25.
39. Goodall, G.J. *et al.* 1983. Production and characterization of antibodies to thymosin beta 4. *J. Immunol.* **131:** 821–825.
40. Cassimeris, L. *et al.* 1992. Thymosin beta 4 sequesters the majority of G-actin in resting human polymorphonuclear leukocytes. *J. Cell Biol.* **119:** 1261–1270.
41. Malinda, K.M. *et al.* 1999. Thymosin beta4 accelerates wound healing. *J. Invest. Dermatol.* **113:** 364–368.
42. Philp, D. *et al.* 2003. Thymosin beta 4 and a synthetic peptide containing its actin-binding domain promote dermal wound repair in db/db diabetic mice and in aged mice. *Wound Repair Regen.* **11:** 19–24.
43. Sosne, G. *et al.* 2001. Thymosin beta 4 promotes corneal wound healing and modulates inflammatory mediators in vivo. *Exp. Eye Res.* **72:** 605–608.
44. Sosne, G. *et al.* 2002. Thymosin beta 4 promotes corneal wound healing and decreases inflammation in vivo following alkali injury. *Exp. Eye Res.* **74:** 293–299.
45. Sosne, G. *et al.* 2005. Thymosin-beta4 modulates corneal matrix metalloproteinase levels and polymorphonuclear cell infiltration after alkali injury. *Invest. Ophthalmol. Vis. Sci.* **46:** 2388–2395.
46. Badamchian, M. *et al.* 2003. Thymosin beta(4) reduces lethality and down-regulates inflammatory mediators in endotoxin-induced septic shock. *Int. Immunopharmacol.* **3:** 1225–1233.
47. Huff, T. *et al.* 2001. Beta-thymosins, small acidic peptides with multiple functions. *Int. J. Biochem. Cell. Biol.* **33:** 205–220.
48. Dunn, S.P. *et al.* 2010. Treatment of chronic nonhealing neurotrophic corneal epithelial defects with thymosin beta 4. *Arch. Ophthalmol.* **128:** 636–638.
49. Allan, C.B. *et al.*, 2011. Effects of Thymosin β4 in a Murine CAESM Model of Experimental Dry Eye (Abstract/Poster). ARVO. Program #: 372011, D915. ARVO, Rockville, MD.
50. Qiu, P. *et al.* 2011. Thymosin beta4 inhibits TNF-alpha-induced NF-kappaB activation, IL-8 expression, and the sensitizing effects by its partners PINCH-1 and ILK. *FASEB J.* **25:** 1815–1826.

Ann. N.Y. Acad. Sci. ISSN 0077-8923

ANNALS OF THE NEW YORK ACADEMY OF SCIENCES
Issue: *Thymosins in Health and Disease*

Neuroprotective and neurorestorative effects of thymosin β4 treatment following experimental traumatic brain injury

Ye Xiong,[1] Asim Mahmood,[1] Yuling Meng,[1] Yanlu Zhang,[1] Zheng Gang Zhang,[2] Daniel C. Morris,[3] and Michael Chopp[2,4]

[1]Departments of Neurosurgery, Henry Ford Health System, Detroit, Michigan. [2]Neurology, Henry Ford Health System, Detroit, Michigan. [3]Emergency Medicine, Henry Ford Health System, Detroit, Michigan. [4]Department of Physics, Oakland University, Rochester, Michigan

Address for correspondence: Ye Xiong, M.D., Ph.D., Henry Ford Health System, Department of Neurosurgery, E&R Building, Room # 3096, 2799 West Grand Boulevard, Detroit, MI 48202. yxiong1@hfhs.org

Traumatic brain injury (TBI) remains a leading cause of mortality and morbidity worldwide. No effective pharmacological treatments are available for TBI because all phase II/III TBI clinical trials have failed. This highlights a compelling need to develop effective treatments for TBI. Endogenous neurorestoration occurs in the brain after TBI, including angiogenesis, neurogenesis, synaptogenesis, oligodendrogenesis, and axonal remodeling, which may be associated with spontaneous functional recovery after TBI. However, the endogenous neurorestoration following TBI is limited. Treatments amplifying these neurorestorative processes may promote functional recovery after TBI. Thymosin beta 4 (Tβ4) is the major G-actin–sequestering molecule in eukaryotic cells. In addition, Tβ4 has other properties including antiapoptosis and anti-inflammation, promotion of angiogenesis, wound healing, stem/progenitor cell differentiation, and cell migration and survival, which provide the scientific foundation for the corneal, dermal, and cardiac wound repair multicenter clinical trials. Here, we describe Tβ4 as a neuroprotective and neurorestorative candidate for treatment of TBI.

Keywords: thymosin beta 4; traumatic brain injury; rat; neuroprotection; neurorestoration

Introduction

An estimated 1.4 million people sustain traumatic brain injury (TBI) each year in the United States, and more than five million people are coping with disabilities from TBI at an annual cost of more than $56 billion.[1] There are no commercially available pharmacological treatment options available for TBI because all clinical trial strategies have failed.[2,3] The disappointing clinical trial results may be due to variability in treatment approaches and heterogeneity of the population of TBI patients.[4–9] Another important aspect is that most clinical trial strategies have used drugs that target a single pathophysiological mechanism, although many mechanisms are involved in secondary injury after TBI.[4] Neuroprotection approaches have historically been dominated by targeting neuron-based injury mechanisms as the primary or even exclusive focus of the neuroprotective strategy.[3] In the vast majority of preclinical studies, the treatment compounds are administered early and, frequently, even before TBI.[10,11] Clinically, the administration of a compound early may be problematic because of the difficulty in obtaining informed consent.[12]

Recent preclinical studies by our group and others have revealed that endogenous neurorestoration is present after TBI, including neurogenesis, axonal sprouting, synaptogenesis, and angiogenesis, which may contribute to the spontaneous functional recovery.[13–18] In addition, treatments that promote these neurorestorative processes have been demonstrated to improve functional recovery after brain injury.[19,20] However, clinical trials in TBI have primarily targeted neuroprotection, and trials directed

doi: 10.1111/j.1749-6632.2012.06683.x

specifically at neurorestoration have not been conducted. The essential difference between neuroprotective and neurorestorative treatments is that the former target the lesion that is still not irreversibly injured and the latter treat the intact tissue.[19] Thus, neurorestorative treatments can be made available for a larger number of TBI patients.

Tβ4 is a multifunctional regenerative small peptide containing 43 amino acids, and it is the major G-actin–sequestering molecule in eukaryotic cells.[21] Tβ4 has prosurvival and proangiogenic properties, protects tissue against damage, and promotes tissue regeneration.[22,23] It also plays a key role in corneal, epidermal, and cardiac wound healing.[21] Tβ4 participates in axonal path finding, neurite formation, cell proliferation, and neuronal survival.[24–26] Our previous studies show that Tβ4 reduces inflammation, stimulates remyelination, and improves functional recovery in animal models of experimental autoimmune encephalomyelitis (EAE) and stroke.[25,27] In summary, these pleiotropic properties make Tβ4 an ideal candidate for treatment of TBI.

Early Tβ4 treatment reduces cortical lesion volume and improves functional recovery after TBI in rats

TBI patients frequently suffer from long-term deficits in cognitive and motor performance. No single animal model can adequately mimic all aspects of human TBI, owing to the heterogeneity of clinical TBI.[11] Some features of cognitive and motor function in humans have been successfully demonstrated in experimental brain trauma models.[28–30] The controlled cortical impact (CCI) model is one of the most widely used TBI models. The CCI-TBI model has many clinically relevant features in that CCI causes not only cortical damage but also selective neuronal death in the hippocampus in rodents, leading to sensorimotor dysfunction and spatial learning and memory deficits, respectively.[18,31–33]

We have evaluated the efficacy of early (6 h postinjury) Tβ4 treatment on spatial learning and sensorimotor functional recovery in rats after TBI induced by unilateral CCI.[34] In brief, TBI rats received Tβ4 at a dose of either 6 or 30 mg/kg or a vehicle control (saline) administered i.p. starting at 6 h after injury and then at 24 and 48 h.[34] Spatial learning was performed during the last five days (31–35 days postinjury) using the modified Morris water maze (MWM) test, which is extremely sensitive to the hippocampal injury.[35–38] Tβ4-treated TBI rats showed significant improvement in spatial learning when compared to the saline-treated TBI rats. Tβ4 treatment also significantly reduced the swim latency to reach the hidden platform by rats post-TBI compared to saline treatment. Using the modified neurological severity score (mNSS) test, our data show that significantly improved scores were observed after TBI in the Tβ4-treated group compared to the saline-treated group. Our data also show that Tβ4 reduced the incidence of both right forelimb and hindlimb footfaults in TBI rats.[34] Histological data show that early Tβ4 treatment reduced cortical lesion volume by 20% and 30% for 6 mg/kg and 30 mg/kg, respectively, and reduced hippocampal cell loss.[34] These findings suggest that TB4 provides neuroprotection even when the treatment was initiated six hours postinjury. In addition, 6-h Tβ4 treatment promotes neurogenesis in the dentate gyrus (DG) of the hippocampus,[38] which may contribute to improvement in spatial learning.

Delayed Tβ4 treatment does not alter cortical lesion volume but improves functional recovery after TBI in rats

The CCI model we used causes cortical tissue loss. Traditionally, the target for neuroprotective treatment of TBI is to reduce the lesion volume.[39,40] A major limitation of neuroprotection strategies is the short time window between injury and treatment. In the vast majority of preclinical TBI studies, the treatment compounds provide neuroprotection only when administered early (usually several hours after brain injury).[11] However, the administration of a compound early in the clinical setting is not practical.[41] The neuroprotective effects demonstrated in rodents may diminish if the treatment compounds are given in the clinical setting beyond the short neuroprotective window. We are able to stimulate recovery of neurological function without altering the lesion volume, which has also been demonstrated in our experimental studies of stroke,[19,42,43] and is, in essence, enhancement of neurorecovery.[19] The extended 24-h window for treatment, which improves neurological recovery, without altering CCI cortical volume, is a major benefit of the neurorestorative therapy.[38] Recently, we evaluated the efficacy of delayed (24 h postinjury) Tβ4 treatment on spatial learning and sensorimotor functional recovery in rats after TBI induced by CCI.[34] Briefly, TBI rats received Tβ4 at a dose of 6 mg/kg or a vehicle (saline) administered i.p. starting 24 h after injury and then every third day for two weeks.[38] The dose of Tβ4 was selected based on our previous studies in animal models of stroke and EAE.[25,27] Tβ4 did not alter lesion volume (14.2 ± 3.9% for saline treatment vs. 15.7 ± 3.6% for Tβ4 treatment).[38] TBI caused neuronal cell loss in the ipsilateral CA3 and DG examined 35 days after injury compared to sham controls. Tβ4 treatment initiated 24 h postinjury significantly reduced cell loss in these two regions compared to saline controls.[38] Tβ4-treated TBI rats showed significant improvement in spatial learning (MWM test) and sensorimotor (mNSS test) functional recovery compared to the saline-treated TBI rats.[38]

Tβ4 treatment promotes neurogenesis after TBI in rats

Evidence accumulated over the past decades has overturned the traditional dogma that the adult mammalian brain cannot generate new neurons. Adult neurogenesis has been identified in all vertebrate species examined thus far, including humans.[44–49] Newly generated neuronal cells originate from neural stem cells in the adult brain. Neural stem cells are the self-renewing, multipotent cells that generate the neuronal and glial cells of the nervous system.[50] The major function of neurogenesis in adult brain seems to replace the neurons that die regularly in certain brain areas. Granule neurons in the DG continuously die, and the progenitors in the subgranular zone of the DG may proliferate at the same rate as mature neuronal death to maintain a constant DG cell number.[51] Similarly, the newly proliferated cells from the subventricular zone migrate and replenish the dead olfactory bulb neurons.[52] Here, we focus on DG neurogenesis, which is important for spatial learning and memory. In normal adult rats, newborn neural cells migrate from the subgranular zone of the DG of the hippocampus into the granule cell layer and eventually become mature granule neurons.[53] These new granule neurons extend axonal processes to their postsynaptic targets[54–57] and receive synaptic input.[58] TBI stimulates widespread cellular proliferation in rats and results in focal neurogenesis in the DG of the hippocampus.[59,60] Some of the newly generated granule neurons integrate into the hippocampus. The integration of the injury-induced neurogenic population into the existing hippocampal circuitry coincides with the time point when cognitive recovery is observed in injured animals.[44]

Bromodeoxyuridine (BrdU), a thymidine analogue, can be incorporated into the DNA of dividing cells and is widely used to label new cells.[61–63] To label proliferating cells, BrdU (100 mg/kg) was injected i.p. daily, starting at day 1 post-TBI for 10 days. The number of $BrdU^+$ cells found in the ipsilateral cortex, DG, and CA3 areas was significantly increased 35 days after TBI compared with sham controls.[18,34,64,65] Tβ4 treatment further increased the number of $BrdU^+$ cells compared to saline controls.[34] The increased number of $BrdU^+$ cells may result from effects of Tβ4 on either increasing cell proliferation or reducing cell death of newborn cells. Our recent data show Tβ4 increases oligodendrocyte precursor cell proliferation and differentiation in animal models of stroke[25] and experimental autoimmune encephalomyelitis.[27] Tβ4 may not directly affect cell proliferation but inhibit cell death, for example, in corneal and conjunctival epithelial cells treated with benzalkonium chloride *in vitro*[66] and endothelial precursor cells under serum deprivation.[67] Our data further show that neurogenesis increases in TBI rats treated with Tβ4,[34,38] suggesting that Tβ4 promotes newborn cells to differentiate into neurons. This is consistent with the effect of Tβ4 on promoting epicardium-derived progenitor cell differentiation into endothelial and smooth muscle cells to form the coronary vasculature.[22] Whether the increased number of $BrdU^+$ cells in the brain of TBI rats treated with Tβ4 is tissue specific remains unknown. Tβ4 may not directly affect cell proliferation. Increased cell proliferation and neurogenesis are also possibly secondary to that of Tβ4-mediated angiogenesis, as described later.

To identify newborn neurons, double immunofluorescent staining for BrdU/NeuN (mature neuronal marker) was performed (Fig. 1). TBI alone significantly increased the number of newborn neurons (NeuN/BrdU-colabeled cells) in the DG of the injured hemisphere.[34,38] Tβ4 treatment significantly further increased the number of newborn neurons compared to saline controls.[34,38] These data suggest that Tβ4 administration initiated 6 or 24 h after TBI promotes neurogenesis in rats.

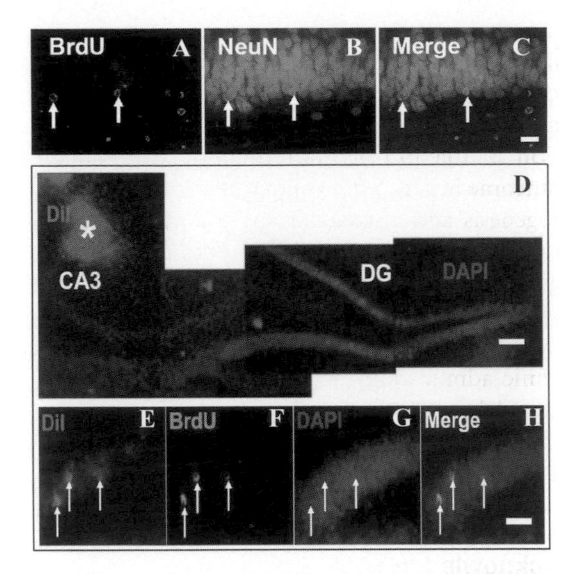

Figure 1. Double immunofluorescent staining for BrdU (red, A) and NeuN (green, B) to identify newborn neurons (yellow after merge, C) in the dentate gyrus of hippocampus from rats examined 35 days after TBI. Micrographs (D) show location of DiI injection in the CA3 region (indicated by white asterisk). In the CA3 region, axons projected from granule neurons in the dentate gyrus will take up injected DiI to their cell bodies. Colocalization (merge, H) of $BrdU^+$ nuclei (green, F) within retrogradely DiI-labeled (red, E) granule cells were examined at 35 days after TBI. Scale bar = 25 μm (C, H). Scale bar = 50 μm (D). Data represent work in progress.

To investigate whether the newborn neurons generated in the DG are capable of projecting their axons into the CA3 region of the hippocampus after TBI, we stereotactically injected a fluorescent tracer, 1,1″-dioleyl-3,3,3″, 3″-tetramethylindocarbocyanine methanesulfonate into the ipsilateral CA3 region (stereotaxic coordinates AP, −3.6 mm bregma, ML, 3.6 mm, DV, 3.0 mm)[86] at day 28 after TBI. BrdU (100 mg/kg, i.p.) was injected i.p. daily starting at day 1 after TBI for 10 days to label newly generated cells. One week after DiI injection (i.e., 35 days after TBI), the animals were anesthetized and sacrificed, and brains were fixed in 4% paraformaldehyde. The brain was cut into seven equally spaced 2-mm coronal blocks using a rat brain matrix. The brain blocks containing the hippocampus were processed for vibratome sections (100 μm) followed by BrdU staining. BrdU and DiI labeling in the hippocampus on brain sections was analyzed with a Bio-Rad MRC 1024 (argon and krypton) laser-scanning confocal imaging system mounted onto a Zeiss microscope. Colocalization of $BrdU^+$ nuclei within retrogradely DiI-labeled granule cells was found, indicating that newborn granule neurons extend axons into the CA3 region that are capable of retrogradely transporting DiI from the CA3 to their cell bodies within the DG after TBI (Fig. 1, preliminary data). This finding suggests that newborn granule neurons may be incorporated into functional hippocampal circuitry after TBI.

Delayed Tβ4 treatment promotes angiogenesis after TBI in rats

The vascular system in the normal adult brain is stable, but is activated in response to certain pathological conditions including injuries.[68] Von Willebrand factor (vWF) staining has been used to identify vascular structure in the brain after TBI.[69] TBI alone significantly increased vascular density in the injured cortex, CA3, and DG of the ipsilateral hemisphere when examined at day 35 after TBI compared to sham controls.[18,34,64,65] Tβ4 treatment significantly increased the vascular density in these regions compared to saline treatment (Fig. 2).[38]

This is in agreement with *in vitro* and *in vivo* proangiogenic effect of Tβ4.[70,71]

Coupling of neurogenesis and angiogenesis

Neurovascular units within the central nervous system consist of endothelial cells, pericytes, neurons, and glial cells, as well as growth factors and extracellular matrix proteins that are close to the endothelium.[72,73] Neurovascular units provide niches for neural stem/progenitor cells in the adult brain and, within these units, newly generated immature neurons are closely associated with the remodeling vasculature. The generation of new vasculature facilitates several coupled neurorestorative processes including neurogenesis and synaptogenesis, which improve functional recovery.[74–76] The vascular production of stromal-derived factor 1 and angiopoietin 1 is involved in neurogenesis and promotes behavioral recovery after stroke.[77] The disruption of this neurovascular coordination has been observed in a variety of brain conditions, including infection, stroke, and trauma.[78] The injured brain promotes angiogenesis and neurogenesis,[13,32,69,79–84] that may contribute to spontaneous functional recovery from injuries such as stroke and TBI. Neurorestorative agents that increase angiogenesis and neurogenesis have been shown to improve

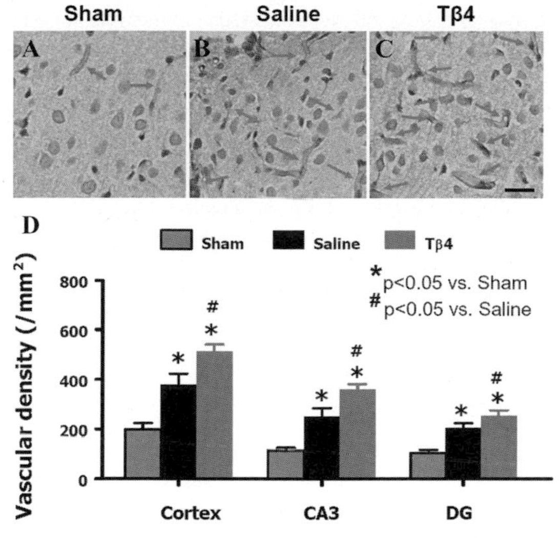

Figure 2. Delayed Tβ4 treatment increases vascular density in the injured cortex, ipsilateral dentate gyrus, and CA3 region 35 days after TBI. Arrows show vWF-stained vascular structure. TBI alone (B) significantly increases the vascular density in the injured cortex compared to sham controls (A) ($P < 0.05$). Tβ4 treatment (C) further enhances angiogenesis after TBI compared to the saline-treated groups ($P < 0.05$). The density of vWF-stained vasculature in different regions is shown in (D). Scale bar = 25 μm (C). Data represent mean ± SD. $*P < 0.05$ versus sham group. $^{\#}P < 0.05$ versus saline group. N (rats/group) = 6 (sham); 9 (saline); and 10 (Tβ4). Data are from Ref. 34.

functional outcome following brain injury.[19,33] Vascular endothelial cells within the neurovascular niche affect neurogenesis directly via contact with neural progenitor cells, while soluble factors from the vascular system that are released into the CNS enhance neurogenesis via paracrine signaling.[85] Here, we demonstrate that Tβ4 treatment promotes both angiogenesis and neurogenesis in rats after TBI,[34,38] suggesting that the neurovascular remodeling at least partially contributes to Tβ4-mediated improvement in functional recovery. A better understanding of molecular mechanisms in the neurovascular niches will be important for developing novel angiogenic and neurogenic therapies for brain injuries.

Conclusion

These previously published studies demonstrate that in the animal model of TBI, early (6 h postinjury) treatment with Tβ4 i.p. at doses of 6 and 30 mg/kg reduces cortical lesion volume and hippocampal cell loss and improves functional recovery, suggesting its potential as a neuroprotective therapy[34] for TBI. More importantly, delayed (24 h postinjury) treatment with Tβ4 administered i.p. at a dose of 6 mg/kg does not reduce lesion volume but significantly improves functional outcome in rats.[38] Tβ4-induced angiogenesis, neurogenesis and oligodendrogenesis may contribute to functional recovery.[38] Therefore, our data suggest that promoting endogenous neurorestorative processes using Tβ4 provides a novel therapeutic option for TBI. It should be noted that systemic administration of Tβ4 is safe and well tolerated by animals and humans.[26] Further investigation of the molecular mechanisms underlying Tβ4-mediated neuroprotection and neurorestoration is warranted.

Acknowledgments

The authors would like to thank Susan MacPhee-Gray for editorial assistance. Tβ4 was provided by RegeneRx Biopharmaceuticals, Inc. under the Materials Transfer Agreement. This work was supported by NIH Grants RO1 NS62002 (Ye Xiong), PO1 NS42345 (Asim Mahmood, Michael Chopp), and PO1 NS023393 (Michael Chopp).

Conflicts of interest

The authors declare no conflicts of interest.

References

1. Langlois, J.A., W. Rutland-Brown & M.M. Wald. 2006. The epidemiology and impact of traumatic brain injury: a brief overview. *J. Head Trauma Rehabil.* **21:** 375–378.
2. Janowitz, T. & D.K. Menon. 2010. Exploring new routes for neuroprotective drug development in traumatic brain injury. *Sci. Transl. Med.* **2:** 27rv21.
3. Loane, D.J. & A.I. Faden. 2010. Neuroprotection for traumatic brain injury: translational challenges and emerging therapeutic strategies. *Trends Pharmacol. Sci.* **31:** 596–604.
4. Stein, D.G. & D.W. Wright. 2010. Progesterone in the clinical treatment of acute traumatic brain injury. *Expert Opin. Investig. Drugs* **19:** 847–857.
5. Aarabi, B. & J.M. Simard. 2009. Traumatic brain injury. *Curr. Opin. Crit. Care* **15:** 548–553.
6. Xiong, Y., A. Mahmood & M. Chopp. 2009. Emerging treatments for traumatic brain injury. *Expert Opin. Emerg. Drugs* **14:** 67–84.
7. Beauchamp, K., H. Mutlak, W.R. Smith, *et al.* 2008. Pharmacology of traumatic brain injury: where is the "golden bullet"? *Mol. Med.* **14:** 731–740.
8. Tolias, C.M. & M.R. Bullock. 2004. Critical appraisal of neuroprotection trials in head injury: what have we learned? *NeuroRx* **1:** 71–79.

9. Narayan, R.K., M. E. Michel, B. Ansell, *et al.* 2002. Clinical trials in head injury. *J. Neurotrauma* **19:** 503–557.
10. Wada, K., K. Chatzipanteli, R. Busto, *et al.* 1998. Role of nitric oxide in traumatic brain injury in the rat. *J. Neurosurg.* **89:** 807–818.
11. Marklund, N. & L. Hillered. 2011. Animal modelling of traumatic brain injury in preclinical drug development: where do we go from here? *Br. J. Pharmacol.* **164:** 1207–1229.
12. Menon, D.K. 2009. Unique challenges in clinical trials in traumatic brain injury. *Crit. Care Med.* **37:** S129–S135.
13. Lu, D., A. Mahmood, C. Qu, *et al.* 2005. Erythropoietin enhances neurogenesis and restores spatial memory in rats after traumatic brain injury. *J. Neurotrauma* **22:** 1011–1017.
14. Lu, D., A. Goussev, J. Chen, *et al.* 2004. Atorvastatin reduces neurological deficit and increases synaptogenesis, angiogenesis, and neuronal survival in rats subjected to traumatic brain injury. *J. Neurotrauma* **21:** 21–32.
15. Zhang, Y., Y. Xiong, A. Mahmood, *et al.* 2010. Sprouting of corticospinal tract axons from the contralateral hemisphere into the denervated side of the spinal cord is associated with functional recovery in adult rat after traumatic brain injury and erythropoietin treatment. *Brain Res.* **1353:** 249–257.
16. Oshima, T., S. Lee, A. Sato, *et al.* 2009. TNF-alpha contributes to axonal sprouting and functional recovery following traumatic brain injury. *Brain Res.* **1290:** 102–110.
17. Richardson, R.M., D. Sun & M.R. Bullock. 2007. Neurogenesis after traumatic brain injury. *Neurosurg. Clin. N. Am.* **18:** 169–181, xi.
18. Xiong, Y., A. Mahmood, Y. Meng, *et al.* 2010. Delayed administration of erythropoietin reducing hippocampal cell loss, enhancing angiogenesis and neurogenesis, and improving functional outcome following traumatic brain injury in rats: comparison of treatment with single and triple dose. *J. Neurosurg.* **113:** 598–608.
19. Zhang, Z.G. & M. Chopp. 2009. Neurorestorative therapies for stroke: underlying mechanisms and translation to the clinic. *Lancet Neurol.* **8:** 491–500.
20. Xiong, Y., A. Mahmood & M. Chopp. 2010. Neurorestorative treatments for traumatic brain injury. *Discov. Med.* **10:** 434–442.
21. Goldstein, A.L., E. Hannappel & H.K. Kleinman. 2005. Thymosin beta4: actin-sequestering protein moonlights to repair injured tissues. *Trends Mol. Med.* **11:** 421–429.
22. Smart, N., C.A. Risebro, A.A. Melville, *et al.* 2007. Thymosin beta4 induces adult epicardial progenitor mobilization and neovascularization. *Nature* **445:** 177–182.
23. Philp, D., T. Huff, Y.S. Gho, *et al.* 2003. The actin binding site on thymosin beta4 promotes angiogenesis. *FASEB. J.* **17:** 2103–2105.
24. Sun, W. & H. Kim. 2007. Neurotrophic roles of the beta-thymosins in the development and regeneration of the nervous system. *Ann. N.Y. Acad. Sci.* **1112:** 210–218.
25. Morris, D.C., M. Chopp, L. Zhang, *et al.* 2010. Thymosin beta4 improves functional neurological outcome in a rat model of embolic stroke. *Neuroscience* **169:** 674–682.
26. Crockford, D., N. Turjman, C. Allan, *et al.* 2010. Thymosin beta4: structure, function, and biological properties supporting current and future clinical applications. *Ann. N.Y. Acad. Sci.* **1194:** 179–189.
27. Zhang, J., Z.G. Zhang, D. Morris, *et al.*2009. Neurological functional recovery after thymosin beta4 treatment in mice with experimental auto encephalomyelitis. *Neuroscience* **164:** 1887–1893.
28. Fox, G.B., L. Fan, R.A. Levasseur, *et al.* 1998. Sustained sensory/motor and cognitive deficits with neuronal apoptosis following controlled cortical impact brain injury in the mouse. *J. Neurotrauma* **15:** 599–614.
29. Fujimoto, S.T., L. Longhi, K.E. Saatman, *et al.* 2004. Motor and cognitive function evaluation following experimental traumatic brain injury. *Neurosci. Biobehav. Rev.* **28:** 365–378.
30. Kline, A.E., J.L. Massucci, D.W. Marion, *et al.* 2002. Attenuation of working memory and spatial acquisition deficits after a delayed and chronic bromocriptine treatment regimen in rats subjected to traumatic brain injury by controlled cortical impact. *J. Neurotrauma* **19:** 415–425.
31. Colicos, M.A., C.E. Dixon & P.K. Dash. 1996. Delayed, selective neuronal death following experimental cortical impact injury in rats: possible role in memory deficits. *Brain Res.* **739:** 111–119.
32. Lu, D., C. Qu, A. Goussev, *et al.* 2007. Statins increase neurogenesis in the dentate gyrus, reduce delayed neuronal death in the hippocampal CA3 region, and improve spatial learning in rat after traumatic brain injury. *J. Neurotrauma* **24:** 1132–1146.
33. Zhang, Y., Y. Xiong, A. Mahmood, *et al.* 2009. Therapeutic effects of erythropoietin on histological and functional outcomes following traumatic brain injury in rats are independent of hematocrit. *Brain Res.* **1294:** 153–164.
34. Xiong, Y., Y. Zhang, A. Mahmood, *et al.* 2012. Neuroprotective and neurorestorative effects of thymosin beta4 treatment initiated 6 hours after traumatic brain injury in rats. *J. Neurosurg.* **116:** 1081–1092.
35. Schallert, T. 2006. Behavioral tests for preclinical intervention assessment. *NeuroRx* **3:** 497–504.
36. Day, L.B. & T. Schallert. 1996. Anticholinergic effects on acquisition of place learning in the Morris water task: spatial mapping deficit or inability to inhibit nonplace strategies? *Behav. Neurosci.* **110:** 998–1005.
37. Choi, S.H., M.T. Woodlee, J.J. Hong, *et al.* 2006. A simple modification of the water maze test to enhance daily detection of spatial memory in rats and mice. *J. Neurosci. Methods* **156:** 182–193.
38. Xiong, Y., A. Mahmood, Y. Meng, *et al.* 2011. Treatment of traumatic brain injury with thymosin beta in rats. *J. Neurosurg.* **114:** 102–115.
39. Cherian, L., J.C. Goodman & C. Robertson. 2007. Neuroprotection with erythropoietin administration following controlled cortical impact injury in rats. *J. Pharmacol. Exp. Ther.* **322:** 789–794.
40. Dietrich, W.D., O. Alonso, R. Busto, *et al.* 1996. Posttreatment with intravenous basic fibroblast growth factor reduces histopathological damage following fluid-percussion brain injury in rats. *J. Neurotrauma* **13:** 309–316.
41. Menon, D.K. & C. Zahed. 2009. Prediction of outcome in severe traumatic brain injury. *Curr. Opin. Crit. Care.* **15:** 437–441.

42. Chen, J. & M. Chopp. 2006. Neurorestorative treatment of stroke: cell and pharmacological approaches. *NeuroRx* **3:** 466–473.
43. Chopp, M. & Y. Li. 2002. Treatment of neural injury with marrow stromal cells. *Lancet Neurol.* **1:** 92–100.
44. Sun, D., M.J. McGinn, Z. Zhou, *et al.* 2007. Anatomical integration of newly generated dentate granule neurons following traumatic brain injury in adult rats and its association to cognitive recovery. *Exp. Neurol.* **204:** 264–272.
45. Kempermann, G. & G. Kronenberg. 2003. Depressed new neurons—adult hippocampal neurogenesis and a cellular plasticity hypothesis of major depression. *Biol. Psychiatry* **54:** 499–503.
46. Kempermann, G. & F.H. Gage. 2002. Genetic influence on phenotypic differentiation in adult hippocampal neurogenesis. *Brain Res. Dev. Brain Res.* **134:** 1–12.
47. van Praag, H., A.F. Schinder, B.R. Christie, *et al.* 2002. Functional neurogenesis in the adult hippocampus. *Nature* **415:** 1030–1034.
48. Hagg, T. 2005. Molecular regulation of adult CNS neurogenesis: an integrated view. *Trends Neurosci.* **28:** 589–595.
49. Richardson, R.M., A. Singh, D. Sun, *et al.* 2010. Stem cell biology in traumatic brain injury: effects of injury and strategies for repair. *J. Neurosurg.* **112:** 1125–1138.
50. Taupin, P. 2006. The therapeutic potential of adult neural stem cells. *Curr. Opin. Mol. Ther.* **8:** 225–231.
51. Gage, F.H. 2000. Mammalian neural stem cells. *Science* **287:** 1433–1438.
52. Yamashima, T., A.B. Tonchev & M. Yukie. 2007. Adult hippocampal neurogenesis in rodents and primates: endogenous, enhanced, and engrafted. *Rev. Neurosci.* **18:** 67–82.
53. Cameron, H.A., C.S. Woolley, B.S. McEwen, *et al.* 1993. Differentiation of newly born neurons and glia in the dentate gyrus of the adult rat. *Neuroscience* **56:** 337–344.
54. Markakis, E.A. & F.H. Gage. 1999. Adult-generated neurons in the dentate gyrus send axonal projections to field CA3 and are surrounded by synaptic vesicles. *J. Comp. Neurol.* **406:** 449–460.
55. Stanfield, B.B. & J.E. Trice. 1988. Evidence that granule cells generated in the dentate gyrus of adult rats extend axonal projections. *Exp. Brain Res.* **72:** 399–406.
56. Hastings, N.B. & E. Gould. 1999. Rapid extension of axons into the CA3 region by adult-generated granule cells. *J. Comp. Neurol.* **413:** 146–154.
57. Hastings, N.B., M.I. Seth, P. Tanapat, *et al.* 2002. Granule neurons generated during development extend divergent axon collaterals to hippocampal area CA3. *J. Comp. Neurol.* **452:** 324–333.
58. Kaplan, M.S. & J.W. Hinds. 1977. Neurogenesis in the adult rat: electron microscopic analysis of light radioautographs. *Science* **197:** 1092–1094.
59. Urrea, C., D.A. Castellanos, J. Sagen, *et al.* 2007. Widespread cellular proliferation and focal neurogenesis after traumatic brain injury in the rat. *Restor. Neurol. Neurosci.* **25:** 65–76.
60. Kernie, S.G., T.M. Erwin & L.F. Parada. 2001. Brain remodeling due to neuronal and astrocytic proliferation after controlled cortical injury in mice. *J. Neurosci. Res.* **66:** 317–326.
61. Taupin, P. 2007. BrdU immunohistochemistry for studying adult neurogenesis: paradigms, pitfalls, limitations, and validation. *Brain Res. Rev.* **53:** 198–214.
62. Taupin, P. 2007. Protocols for studying adult neurogenesis: insights and recent developments. *Regen. Med.* **2:** 51–62.
63. Kuhn, H.G. & C.M. Cooper-Kuhn. 2007. Bromodeoxyuridine and the detection of neurogenesis. *Curr. Pharm. Biotechnol.* **8:** 127–131.
64. Meng, Y., Y. Xiong, A. Mahmood, *et al.* 2011. Dose-dependent neurorestorative effects of delayed treatment of traumatic brain injury with recombinant human erythropoietin in rats. *J. Neurosurg.* **115:** 550–560.
65. Ning, R., Y. Xiong, A. Mahmood, *et al.* 2011. Erythropoietin promotes neurovascular remodeling and long-term functional recovery in rats following traumatic brain injury. *Brain Res.* **1384:** 140–150.
66. Sosne, G., A.R. Albeiruti, B. Hollis, *et al.* 2006. Thymosin beta4 inhibits benzalkonium chloride-mediated apoptosis in corneal and conjunctival epithelial cells *in vitro*. *Exp. Eye Res.* **83:** 502–507.
67. Zhao, Y., F. Qiu, S. Xu, *et al.* 2011. Thymosin beta4 activates integrin-linked kinase and decreases endothelial progenitor cells apoptosis under serum deprivation. *J. Cell Physiol.* **226:** 2798–2806.
68. Greenberg, D.A. & K. Jin. 2005. From angiogenesis to neuropathology. *Nature* **438:** 954–959.
69. Lu, D., A. Mahmood, R. Zhang, *et al.* 2003. Upregulation of neurogenesis and reduction in functional deficits following administration of DEtA/NONOate, a nitric oxide donor, after traumatic brain injury in rats. *J. Neurosurg.* **99:** 351–361.
70. Smart, N., A. Rossdeutsch & P.R. Riley. 2007. Thymosin beta4 and angiogenesis: modes of action and therapeutic potential. *Angiogenesis* **10:** 229–241.
71. Dettin, M., F. Ghezzo, M.T. Conconi, *et al.* 2011. *In vitro* and *in vivo* pro-angiogenic effects of thymosin-beta4-derived peptides. *Cell Immunol.* **271:** 299–307.
72. Lok, J., P. Gupta, S. Guo, *et al.* 2007. Cell-cell signaling in the neurovascular unit. *Neurochem. Res.* **32:** 2032–2045.
73. Guo, S. & E.H. Lo. 2009. Dysfunctional cell-cell signaling in the neurovascular unit as a paradigm for central nervous system disease. *Stroke* **40:** S4–S7.
74. Beck, H. & K.H. Plate. 2009. Angiogenesis after cerebral ischemia. *Acta Neuropathol.* **117:** 481–496.
75. Chopp, M. & Y. Li. 2008. Treatment of stroke and intracerebral hemorrhage with cellular and pharmacological restorative therapies. *Acta. Neurochir. Suppl.* **105:** 79–83.
76. Li, Y. & M. Chopp. 2009. Marrow stromal cell transplantation in stroke and traumatic brain injury. *Neurosci. Lett.* **456:** 120–123.
77. Ohab, J.J., S. Fleming, A. Blesch, *et al.* 2006. A neurovascular niche for neurogenesis after stroke. *J. Neurosci.* **26:** 13007–13016.
78. Han, H.S. & K. Suk. 2005. The function and integrity of the neurovascular unit rests upon the integration of the vascular and inflammatory cell systems. *Curr. Neurovasc. Res.* **2:** 409–423.
79. Chopp, M., Y. Li & Z.G. Zhang. 2009. Mechanisms underlying improved recovery of neurological function after stroke

in the rodent after treatment with neurorestorative cell-based therapies. *Stroke* **40:** S143–S145.
80. Mahmood, A., D. Lu & M. Chopp. 2004. Marrow stromal cell transplantation after traumatic brain injury promotes cellular proliferation within the brain. *Neurosurgery* **55:** 1185–1193.
81. Chopp, M., Y. Li & J. Zhang. 2008. Plasticity and remodeling of brain. *J. Neurol. Sci.* **265:** 97–101.
82. Wang, L., Z. Zhang, Y. Wang, *et al.* 2004. Treatment of stroke with erythropoietin enhances neurogenesis and angiogenesis and improves neurological function in rats. *Stroke* **35:** 1732–1737.
83. Xiong, Y., A. Mahmood, D. Lu, *et al.* 2008. Histological and functional outcomes after traumatic brain injury in mice null for the erythropoietin receptor in the central nervous system. *Brain Res.* **1230:** 247–257.
84. Xiong, Y., D. Lu, C. Qu, *et al.* 2008. Effects of erythropoietin on reducing brain damage and improving functional outcome after traumatic brain injury in mice. *J. Neurosurg.* **109:** 510–521.
85. Yang, X.T., Y.Y. Bi & D.F. Feng. 2011. From the vascular microenvironment to neurogenesis. *Brain Res. Bull.* **84:** 1–7.
86. Paxinos, G. & C. Watson. 1997. The Rat Brain in Stereotactic Coordinates. Academic Press, Inc. San Diego.

Ann. N.Y. Acad. Sci. ISSN 0077-8923

ANNALS OF THE NEW YORK ACADEMY OF SCIENCES
Issue: *Thymosins in Health and Disease*

Use of the cardioprotectants thymosin β4 and dexrazoxane during congenital heart surgery: proposal for a randomized, double-blind, clinical trial

Daniel Stromberg,[1] Tia Raymond,[1] David Samuel,[1] David Crockford,[2] William Stigall,[1] Steven Leonard,[3] Eric Mendeloff,[1] and Andrew Gormley[4]

[1]Congenital Heart Surgery Unit, Medical City Children's Hospital, Dallas, Texas. [2]RegeneRx Biopharmaceuticals, Inc., Rockville, Maryland. [3]Congenital Heart Surgery, Rocky Mountain Hospital for Children, Denver, Colorado. [4]Division of Critical Care, University of Oklahoma Children's Hospital, Oklahoma City, Oklahoma

Address for correspondence: Daniel Stromberg, M.D., Medical City Children's Hospital, 7777 Forest Lane, Suite B-246, Dallas, TX 75230. dstrom115@yahoo.com

Neonates and infants undergoing heart surgery with cardioplegic arrest experience both inflammation and myocardial ischemia-reperfusion (IR) injury. These processes provoke myocardial apoptosis and oxygen-free radical formation that result in cardiac injury and dysfunction. Thymosin β4 (Tβ4) is a naturally occurring peptide that has cardioprotective and antiapoptotic effects. Similarly, dexrazoxane provides cardioprotection by reduction of toxic reactive oxygen species (ROS) and suppression of apoptosis. We propose a pilot pharmacokinetic/safety trial of Tβ4 and dexrazoxane in children less than one year of age, followed by a randomized, double-blind, clinical trial of Tβ4 or dexrazoxane versus placebo during congenital heart surgery. We will evaluate postoperative time to resolution of organ failure, development of low cardiac output syndrome, length of cardiac ICU and hospital stays, and echocardiographic indices of cardiac dysfunction. Results could establish the clinical utility of Tβ4 and/or dexrazoxane in ameliorating ischemia-reperfusion injury during congenital heart surgery.

Keywords: thymosin β4; dexrazoxane; congenital heart surgery; cardiopulmonary bypass

Introduction

The deleterious effects of cardiopulmonary bypass (CPB) with cardioplegic arrest of the heart during congenital heart operations greatly influence postoperative morbidity and mortality. Neonates and infants undergoing cardiac surgery experience both a systemic inflammatory response, and myocardial IR injury as cardioplegic arrest is reversed. These processes provoke elaboration of cytokines and activation of the complement cascade, as well as oxygen free radical formation and induction of myocardial apoptosis.[1–3] Frequently, myocardial injury and cardiac dysfunction ensue, leading to low cardiac output syndrome and multisystem organ failure. The irreversible component of these injuries, in addition to the abnormal workloads imposed on the myocardium from the anatomic defects themselves, may have consequences for long-term cardiac function, and may, in part, explain contractile dysfunction observed late after congenital heart surgery.

In an attempt to mitigate the severity of the inflammatory response seen after CPB, many centers administer steroids prior to or during cardiac surgery. Though controversial,[4–6] several investigations have supported these practices based upon their ability to improve inflammatory profiles and early postoperative clinical indices.[7–9] In contrast, few therapies have been clinically tested for alleviation of the myocardial IR injury inherent to the process of complex newborn heart surgery. Clancey *et al.* evaluated the neurocardiac protective effects of allopurinol in infants undergoing deep hypothermic circulatory arrest. They found that allopurinol reduced the incidence of a composite variable of

doi: 10.1111/j.1749-6632.2012.06710.x

death, seizures, coma, or cardiac events, but only in hypoplastic left heart syndrome (HLHS) infants—non-HLHS infants did not benefit.[10] Jin *et al.* evaluated pre-CPB use of adenosine infusion to reduce IR injury during congenital heart surgery. They demonstrated decreased troponin I at six and 24 h postoperatively, and a decreased ICU length of stay compared to control subjects. However, the authors noted that they have abandoned the practice of preoperative adenosine infusion due to frequent hypotension in recipients.[11] In adults, deferoxamine, an iron chelator and ROS scavenger, has been investigated.[12–14] However, it has not proven to be clinically useful, perhaps because it is unable to penetrate the cell membrane to ameliorate ROS-induced intracellular injury.

Thymosin β4 (Tβ4), a naturally occurring 43 amino acid peptide, is a multifunctional protein expressed in the developing heart, platelets, and white cells.[15] It affects actin–cytoskeleton organization necessary for cell motility and organogenesis, and it promotes skin and corneal wound healing in humans.[16–18] Recently, Tβ4 has been shown to be cardioprotective in animal models of myocardial ischemic injury. It enhances cardiomyocyte survival through antiapoptotic activation of integrin-linked kinase and Akt.[15] It also promotes stem cell activation, migration, and differentiation, as well as coronary revascularization.[19] Systemic administration of Tβ4 in cardiac injury has reduced apoptotic cell death as assessed by TUNEL assay, decreased myocardial scar volume and fibrosis, improved postinfarction cardiac function, and enhanced cell survival.[15] Furthermore, Tβ4 is known to reduce inflammation by blocking NF-κB activation and upregulating IL-10, and it possesses antimicrobial properties.[16] The peptide's safety and pharmacokinetics during systemic administration in man have recently been established.[20]

Dexrazoxane is a cyclic derivative of EDTA that is FDA approved for the prevention of anthracycline-related cardiotoxicity in humans. It is cell membrane permeable, and functions by binding low molecular weight ferrous iron in the intracellular space, which reduces the formation of anthracycline–iron complexes and the consequent generation of reactive oxygen species toxic to cardiomyocytes.[21] It may also exert protective effects via iron-independent free radical scavenging,[22] suppression of apoptotic signaling,[23–25] and inhibition of DNA topoisomerase II.[26] Dexrazoxane is not protein bound in serum, and has a distribution half-life of 12–60 minutes.[27] It is excreted in the kidney, with 40–60% of an i.v. dose excreted into the urine within 24 hours.[28]

Dexrazoxane has been studied extensively in children receiving anthracycline chemotherapy. Lipschultz *et al.* showed in a prospective randomized clinical trial that dexrazoxane prevents or reduces myocardial injury as reflected by decreased elevation of troponin T.[29] Elbl *et al.* found that dexrazoxane reduced the risk of clinical and subclinical anthracycline cardiotoxicity over a seven-year follow-up period (without affect upon the response rate to chemotherapy) via echocardiographic evaluation.[30] Several more recent prospective investigations have corroborated dexrazoxane's cardioprotective effects in conjunction with chemotherapy.[31,32] No adverse reactions to dexrazoxane have been observed in pediatric trials, save for a report concerned with an association between dexrazoxane and secondary malignant neoplasms in children with Hodgkin's lymphoma who received high doses of three chemotherapeutic DNA topoisomerase II inhibitors.[33] However, the chemotherapeutic drugs used were themselves known to be associated with secondary malignancies in childhood cancer survivors. Furthermore, several more recent, large, pediatric case series involving oncology patients have not found a link between dexrazoxane and malignancy.[34–36]

Hypotheses: cardioprotection

Given the information above, we hypothesize that perioperative administration of dexrazoxane or Tβ4 during neonatal/infant congenital heart surgery will result in:

1. A 20% decrease in time to resolution of organ failure—the primary outcome variable—defined as days to the point of being off invasive mechanical ventilation, off renal replacement therapy, and off significant inotropic support (defined as milrinone >0.5 mcg/kg/min, dopamine >3 mcg/kg/min, dobutamine >3 mcg/kg/min or any combination of these inotropes, or any epinephrine, norepinephrine, phenylephrine or vasopressin). One point will be awarded for each postoperative day of continued organ dysfunction up to postoperative

day 14. A score of 15 will be assigned if organ failure is not resolved by postoperative day 14, or if the patient requires mechanical circulatory support or experiences mortality. This variable has been chosen to allow for recognition of early drug effects, and those that might be delayed beyond the immediate postoperative period.

2. Decreased incidence of postoperative low cardiac output syndrome (LCOS)—binary variable as defined by Hoffmann *et al.*:[37] clinical signs or symptoms of LCOS (tachycardia, oliguria, poor perfusion, cardiac arrest) requiring either mechanical support, an increase in the amount of inotropic support >100% above baseline at CICU admission, administration of a new inotrope, or initiation of pacing specifically to treat LCOS.
3. Decreased myocardial injury—as determined by elevated serum cardiac troponin T (>0.01 ng/mL), and severely elevated cardiac troponin T (>0.025 ng/mL),[29] drawn on postoperative days 1 and 7.
4. Decreased oxidative stress—as measured by lipoperoxidation (urine F2-isoprostane), and urine total antioxidant activity (TAS via Calbiochem assay) on days 0, 1, 3, and 7.
5. Reduced inflammatory activation—IL-6 and IL-10 will be measured at time 0, 12, and 24 h postoperatively.
6. Decreased early and intermediate myocardial dysfunction—as measured by the Tei index, tissue Doppler E/E′ ratio, and ventricular ejection fraction on postoperative days 1 and 7, and at days 30–60.
7. Decreased length of ICU and overall hospital stays.

Hypothesis: neuroprotection

It is known that cerebral injury occurs by multiple mechanisms in children with congenital heart disease who undergo surgical intervention. These mechanisms include hypoxic-ischemic injury secondary to hypoperfusion, deep hypothermic circulatory arrest, or air embolization; systemic inflammation; excitatory neurotransmitter release; and oxidative stress.[38] Given the ability of dexrazoxane to oppose norepinephrine vascular contraction via alpha-2 adrenoreceptor-mediated release of relaxing factors,[39] to scavenge ROS, and to inhibit apoptosis, it has been suggested that dexrazoxane might mitigate neurologic IR injury during congenital heart operations. Similarly, Tβ4, which is induced in the brain following ischemia, has also been found to possess neuroprotective properties. Popoli *et al.* demonstrated that Tβ4 prevented loss of rat hippocampal neurons after kainic acid treatment.[40] Zhang *et al.* showed improved functional recovery of mice with experimental autoimmune encephalitis after intraperitoneal injection of Tβ4.[41]

Activin A is a glycoprotein upregulated in the presence of perioperative neurologic injury in infants which serves as a marker of brain insult.[42] We hypothesize that administration of either dexrazoxane or Tβ4 in the perioperative period will decrease neurologic injury during neonatal/infant congenital heart surgery. This will be assessed by both urine activin A concentration measured preoperatively, and on day 0 (end of CPB and on CICU admission), and day 1 (at 12 and 24 h postoperatively); and clinically by physical examination and assignment of a Pediatric Cerebral Performance Category score[43] at hospital discharge.

Hypothesis: safety and pharmacokinetics

Although both dexrazoxane and Tβ4 have been used in human trials without adverse outcome, to our knowledge, neither of these medications has been studied in human neonates and infants. We hypothesize that each medication may be safely administered to neonates and infants in the perioperative period without significant morbidity or increase in mortality. Thus, we propose a pretrial pilot study of each medication in a population of children deemed most at risk for adverse drug reaction, namely children under four months of age. This pilot investigation will establish the safety profiles and pharmacokinetics of these drugs in children undergoing congenital heart surgery. Perioperative complications/morbidities and mortality will be recorded throughout the investigation using the Common Terminology Criteria for Adverse Events (v4.03, June 14, 2010), and compared between groups. A Data Safety and Monitoring Board (DSMB) will be established to monitor patient outcomes at the outset, midpoint, and conclusion of the study. Pilot results could inform decisions regarding study drug dosage and timing of drug administration in the larger trial.

Study design

Pretrial pilot

This will be a prospective, blinded, single-center, safety/pharmacokinetic trial of children up to four months of age who will be randomized for enrollment in either the Tβ4 or dexrazoxane groups. Single ventricle patients will be excluded. For each medication, we will test three dosing regimens: Tβ4 at 5 mg/kg/dose, 12.5 mg/kg/dose, and 20 mg/kg/dose,[20] and dexrazoxane at 200 mg/m^2/dose, 300 mg/m^2/dose, 400 mg/m^2/dose.[44] Tβ4 and dexrazoxane will be administered IV via a central venous catheter. Both medications will be given in the operating room 15–30 minutes prior to CPB (dose #1), after ultrafiltration/separation from bypass (dose #2), and on the morning following surgery in the ICU (dose #3).

Four patients will be assigned to each experimental dosing regimen, for a total of 12 patients studied per drug. Serum samples will be obtained for pharmacokinetic testing at five minutes after drug infusion on the morning following surgery (dose #3), three hours later, and at the theoretical complete elimination time of 24 h after administration of dose #3. The blood volume required for testing at each time point will be 0.5 mL per draw.

Adverse events following drug administration will be recorded using the National Cancer Institute Common Terminology Criteria for Adverse Events (v4.03, June 14, 2010). Since the study population represents a high-risk group of patients *a priori*, subjects will be assigned an illness severity score preoperatively (calculated on the day of surgery—one point given for inpatient status, PGE infusion, O2 saturation <90%, mechanical ventilation, inotropic support).[45] Complications will be compared to those encountered in similar age/diagnosis/illness severity matched patients per programmatic morbidity and mortality records.

Stopping rules for drug treatment will include (1) hemodynamic instability immediately before study drug administration (as determined by the attending cardiovascular anesthesiologist or surgeon); (2) occurrence of one or more adverse events assessed as severe, and probably or definitely related to study drug administration; or (3) occurrence of three or more serious, non-life-threatening adverse events assessed as probably or definitely related to study drug administration.

All adverse events will be reported to the DSMB and evaluated before proceeding with the larger randomized, clinical trial.

Main investigation

The main investigation will be a prospective, randomized, double-blind, multicenter clinical trial, with block randomization for single ventricle patients. Participating centers shall collaborate through the HCA Pediatric Critical Care Research Consortium after site-specific Institutional Review Board approval. Medical City Children's Hospital, Dallas will serve as the data coordinating center. Inclusion criteria for the investigation shall be (1) age ≤ 1 year, (2) open heart surgery requiring CPB and use of cardioplegia, and (3) parent/guardian consent for the study obtained. Exclusion criteria will be (1) gestational age < 36 weeks, (2) known syndrome or genetic abnormality, except trisomy 21, (3) concurrent enrollment in another research protocol, and (4) no parental/guardian consent obtained.

Treatment protocol

After diagnostic information and preoperative illness severity scores are assigned, consecutive patients less than one year of age will be randomized to receive saline placebo, dexrazoxane 300 mg/m^2/dose i.v., or Tβ4 20 mg/kg/dose i.v. (final dose determination to be informed by pilot investigation). The study drug will be administered prior to CPB, after ultrafiltration/separation from CPB, and on the morning following surgery. Intraoperative variables including CPB time, deep hypothermic circulatory arrest time, selective cerebral protection time, and aortic cross clamp time will be recorded. Myocardial injury will be assessed by serial troponin T assays conducted preoperatively (day 0), and on postoperative days 1, 3, and 7. Oxidative stress will be determined via urinary F2-isoprostane and plasma TAS assays on days 0, 1, 3, and 7. Inflammatory profiles will be evaluated by IL-6 and IL-10 levels preoperatively, and at 12 and 24 h postoperatively. Early heart function will be examined by echocardiography on postoperative days 1 and 7; intermediate function will be evaluated between days 30 and 60. Neurologic injury will be assessed by urine activin A concentration obtained preoperatively, at the conclusion of CPB, upon CICU admission, and at 12 and 24 h postoperatively. Experimental groups will be compared preoperatively by age, weight, sex, diagnosis, and illness severity score calculated on the day of surgery as previously described.[45] Intraoperative variables will also be compared. Postoperative clinical outcomes will be evaluated by assessment of complications, mortality, time to resolution of organ failure, echocardiographic parameters, laboratory values, pediatric cerebral performance category scores at hospital discharge, and cardiac ICU/hospital length of stay. All caregivers will be blinded to patient group assignment. Single ventricle patients will be block randomized to achieve equal numbers in each group.

Statistical analysis

The resolution of organ failure score will be recorded as the primary outcome variable, and intention to treat analysis will be utilized. Assuming a 20% difference in the resolution of organ failure score compared to control, 132 patients will be required for each arm of the study (assuming $\beta = 0.20$, $\alpha = 0.05$). Thus, we will enroll 140 subjects per group, for a total enrollment of 420 patients.

Comparisons between groups will be made using Student's t-test for continuous variables, and chi-square analysis for categorical variables. Multiple-linear regression will be used to investigate effects of several factors simultaneously upon outcome variables.

Safety monitoring

A Data Safety and Monitoring Board will be convened to evaluate pilot data, interim results of the investigation, and concluding data. Liver function tests will be recorded per postoperative clinical routine and analyzed. Similarly, complete blood counts will be scrutinized for evidence of bone marrow suppression.[44] If a clear clinical disadvantage accrues to subjects in one of the experimental groups, then the study will be terminated and results will be publicized. Such surveillance will insure that this investigation conforms to the highest ethical standards.

Feasibility

Based upon the number of congenital heart cases performed each year at centers participating in the HCA Pediatric Critical Care Research Consortium (approximately 800/year), we expect the investigation to require two years to complete enrollment. Data analysis should be completed within six months thereafter. The budget for this study is currently under development.

Significance

Cardioprotective medicines are sorely needed for alleviation of inflammation and IR injury after congenital heart surgery, as these maladies cause significant perioperative morbidity and mortality. Tβ4 and dexrazoxane are excellent candidates for study, as they both possess anti-inflammatory and antiapoptotic properties that work through different, and potentially complimentary, cellular mechanisms. Tβ4 decreases inflammation through its effect upon NF-κB; dexrazoxane does so by topoisomerase inhibition. Tβ4 suppresses apoptosis via effect upon Akt cell survival pathways; dexrazoxane reduces the ROS stimulus to apoptotic signaling.

While the known safety profiles for both medicines are encouraging, this study will further evaluate their safety and pharmacokinetics in a neonatal (i.e., vulnerable) population prior to initiation of the larger trial. A DSMB will be utilized throughout to ensure ethical conduct and investigative efficiency.

Conclusive data from a well-designed, double-blind, randomized clinical trial should provide the proof-of-concept regarding antiapoptotic therapy in the management of perioperative heart failure. Moreover, secondary outcome measure analysis relative to neurologic injury markers might inform additional studies on perioperative neuroprotection. Ultimately, our goal is to identify cardioprotective medicines that may improve clinical outcomes for children undergoing congenital heart surgery.

Conflicts of interest

The authors declare no conflicts of interest.

References

1. Chaney, M.A. 2002. Corticosteroids and cardiopulmonary bypass: a review of clinical investigations. *Chest* **121:** 921–931.
2. Caputo, M., A. Mokhtari, C.A. Rogers, *et al.* 2009. The effects of normoxic versus hyperoxic cardiopulmonary bypass on oxidative stress and inflammatory response in cyanotic pediatric patients undergoing open cardiac surgery: a randomized controlled trial. *J. Thorac. Cardiovasc. Surg.* **138:** 206–214.
3. Hare, J.M. 2001. Oxidative stress and apoptosis in heart failure progression. *Circ. Res.* **89:** 198–200.
4. Pasquali, S.K., M. Hall, J.S. Li, *et al.* 2010. Corticosteroids and outcome in children undergoing congenital heart surgery: analysis of the pediatric health information systems database. *Circulation* **122:** 2123–2130.

5. Robertson-Malt, S., B. Afrane & M. Elbarbary. 2009. Prophylactic steroids for pediatric open heart surgery (Review). *Cochrane Collaboration* **1:** 1–24.
6. Graham, E.M., A.M. Atz, R.J. Butts, *et al.* 2011. Standardized preoperative corticosteroid treatment in neonates undergoing cardiac surgery: results from a randomized trial. *J. Thorac. Cardiovasc. Surg.* **142:** 1523–1529.
7. Schroeder, V.A., J.M. Pearl, S.M. Schwartz, *et al.* 2003. Combined steroid treatment for congenital heart surgery improves oxygen delivery and reduces postbypass inflammatory mediator expression. *Circulation* **107:** 2823–2828.
8. Checchia, P.A., C.L. Backer, R.A. Bronicki, *et al.* 2003. Dexamethasone reduces postoperative troponin levels in children undergoing cardiopulmonary bypass. *Crit. Care Med.* **31:** 1742–1745.
9. Clarizia, N.A., C. Manlhiot, S.M. Schwartz, *et al.* 2011. Improved outcomes associated with intraoperative steroid use in high-risk pediatric cardiac surgery. *Ann. Thorac. Surg.* **91:** 1222–1227.
10. Clancy, R.R., S.A. McGaurn, J.E. Goin, *et al.* 2001. Allopurinol neurocardiac protection trial in infants undergoing heart surgery using deep hypothermic circulatory arrest. *Pediatrics* **108:** 61–70.
11. Jin, Z., W. Duan, M. Chen, *et al.* 2011. The myocardial protective effects of adenosine pretreatment in children undergoing cardiac surgery: a randomized controlled clinical trial. *Eur. J. Cardiothor. Surg.* **39:** e90–e96.
12. Ferreira, R., M. Burgos, J. Milei, *et al.* 1990. Effect of supplementing cardioplegic solution with deferoxamine on reperfused human myocardium. *J. Thorac. Cardiovasc. Surg.* **100:** 708–714.
13. Menasché, P., C. Pasquier, S. Bellucci, *et al.* 1988. Deferoxamine reduces neutrophil-mediated free radical production during cardiopulmonary bypass in man. *J. Thorac. Cardiovasc. Surg.* **96:** 582–589.
14. Menasché, P., H. Antebi, L.G. Alcindor, *et al.* 1990. Iron chelation by deferoxamine inhibits lipid peroxidation during cardiopulmonary bypass in humans. *Circulation* **82:** 390–396.
15. Bock-Marquette, I., A. Saxena, M.D. White, *et al.* 2004. Thymosin beta4 activates integrin-linked kinase and promotes cardiac cell migration, survival and cardiac repair. *Nature* **432:** 466–472.
16. Crockford, D., N. Turjman, C. Allan & J. Angel. 2010. Thymosin beta4: structure, function, and biological properties supporting current and future clinical applications. *Ann. N.Y. Acad. Sci.* **1194:** 179–189.
17. Malinda, K.M., G.S. Sidhu, H. Mani, *et al.* 1999. Thymosin beta4 accelerates wound healing. *J. Invest. Dermatol.* **113:** 364–368.
18. Sosne, G., E.A. Szliter, R. Barrett, *et al.* 2002. Thymosin beta 4 promotes corneal wound healing and decreases inflammation in vivo following alkali injury. *Exp. Eye Res.* **74:** 293–299.
19. Shrivastava, S., D. Srivastava, E.N. Olson, *et al.* 2010. Thymosin beta4 and cardiac repair. *Ann. N.Y. Acad. Sci.* **1194:** 87–96.
20. Ruff, D., D. Crockford, G. Girardi & Y. Zhang. 2010. A randomized, placebo-controlled, single and multiple dose study of intravenous thymosin beta4 in healthy volunteers. *Ann. N.Y. Acad. Sci.* **1194:** 223–229.
21. Zheng, H., C. Dimayuga, A. Hudaihed & S.D. Katz. 2002. Effect of dexrazoxane on homocysteine-induced endothelial dysfunction in normal subjects. *Arterioscler Thromb. Vasc. Biol.* **22:** e15–e18.
22. Junjing, Z., Z. Yan & Z. Baolu. 2010. Scavenging effects of dexrazoxane on free radicals. *J. Clin. Biochem. Nutr.* **47:** 238–245.
23. Popelová, O., M. Sterba, P. Hasková, *et al.* 2009. Dexrazoxane-afforded protection against chronic anthracycline cardiotoxicity in vivo: effective rescue of cardiomyocytes from apoptotic cell death. *Br. J. Cancer* **101:** 792–802.
24. Zhou, L., R.Y.T. Sung, K. Li, *et al.* 2011. Cardioprotective effect of dexrazoxane in a rat model of myocardial infarction: anti-apoptosis and promoting angiogenesis. *Int. J. Cardiol.* **152:** 196–201.
25. Spagnuolo, R.D., S. Recalcati, L. Tacchini & G. Cairo. 2011. Role of hypoxia-inducible factors in the dexrazoxane-mediated protection of cardiomyocytes from doxorubicin-induced toxicity. *Br. J. Pharmacol.* **163:** 299–312.
26. Hasinoff, B.B., P.E. Schroeder & D. Patel. 2003. The metabolites of the cardioprotective drug dexrazoxane do not protect myocytes from doxorubicin-induced cytotoxicity. *Mol. Pharmacol.* **64:** 670–678.
27. Wiseman, L.R. & C.M. Spencer. 1998. Dexrazoxane. A review of its use as a cardioprotective agent in patients receiving anthracycline-based chemotherapy. *Drugs* **56:** 385–403.
28. Brier, M.E., S.K. Gaylor, J.P. McGovren, *et al.* 2011. Pharmacokinetics of dexrazoxane in subjects with impaired kidney function. *J. Clin. Pharmacol.* **51:** 731–738.
29. Lipshultz, S.E., N. Rifai, V.M. Dalton, *et al.* 2004. The effect of dexrazoxane on myocardial injury in doxorubicin-treated children with acute lymphoblastic leukemia. *N. Engl. J. Med.* **351:** 145–153.
30. Elbl, L., H. Hrstkova, I. Tomaskova & J. Michalek. 2006. Late anthracycline cardiotoxicity protection by dexrazoxane (ICRF-187) in pediatric patients: echocardiographic follow-up. *Support Care Cancer* **14:** 128–136.
31. Sanchez-Medina, J., O. Gonzalez-Ramella & S. Gallegos-Castorena. 2010. The effect of dexrazoxane for clinical and subclinical cardiotoxicity in children with acute myeloid leukemia. *J. Pediatr. Hematol. Oncol.* **32:** 294–297.
32. Choi, H.S., E.S. Park, H.J. Kang, *et al.* 2010. Dexrazoxane for preventing anthracycline cardiotoxicity in children with solid tumors. *J. Korean Med. Sci.* **25:** 1336–1342.
33. Tebbi, C.K., W.B. London, D. Friedman, *et al.* 2007. Dexrazoxane-associated risk for acute myeloid leukemia/myelodysplastic syndrome and other secondary malignancies in pediatric Hodgkin's disease. *J. Clin Oncol.* **25:** 493–500.
34. Barry, E.V., L.M. Vrooman, S.E. Dahlberg, *et al.* 2008. Absence of secondary malignant neoplasms in children with high-risk acute lymphoblastic leukemia treated with dexrazoxane. *J. Clin. Oncol.* **26:** 1106–1111.
35. Lipshultz, S.E., R.E. Scully, S.R. Lipsitz, *et al.* 2010. Assessment of dexrazoxane as a cardioprotectant in doxorubicin-treated children with high-risk acute lymphoblastic

leukemia: long-term follow-up of a prospective, randomized, multicentre trial. *Lancet Oncol.* **11:** 950–961.

36. Vrooman, L.M., D.S. Neuberg, K.E. Stevenson, *et al.* 2011. The low incidence of secondary acute myelogenous leukemia in children and adolescents treated with dexrazoxane for acute lymphoblastic leukemia: a report from the Dana-Farber Cancer Institute ALL Consortium. *Eur. J. Cancer* **47:** 1373–1379.
37. Hoffman, T.M., G. Wernovsky, A.M. Atz, *et al.* 2003. Efficacy and safety of milrinone in preventing low cardiac output syndrome in infants and children after corrective surgery for congenital heart disease. *Circulation* **107:** 996–1002.
38. Su, X.W. & A. Undar. 2010. Brain protection during pediatric cardiopulmonary bypass. *Artif. Organs* **34:** E91–E102.
39. Vidrio, H., O.F. Carrasco & R. Rodriguez. 2006. Antivasoconstrictor effect of the neuroprotective agent dexrazoxane in rat aorta. *Life Sci.* **80:** 98–104.
40. Popoli, P., R. Pepponi, A. Martire, *et al.* 2007. Neuroprotective effects of thymosin beta4 in experimental models of excitotoxicity. *Ann. N.Y. Acad. Sci.* **1112:** 219–224.
41. Zhang, J., Z.G. Zhang, D. Morris, *et al.* 2009. Neurological functional recovery after thymosin beta4 treatment in mice with experimental auto encephalomyelitis. *Neuroscience* **164:** 1887–1893.
42. Florio, P., R.F. Abella, T. de la Torre, *et al.* 2007. Perioperative activin A concentrations as a predictive marker of neurologic abnormalities in children after open heart surgery. *Clin. Chem.* **53:** 982–985.
43. Fiser, D.H. 1992. Assessing the outcome of pediatric intensive care. *J. Pediatr.* **121:** 68–74.
44. Holcenberg, J.S., K.D. Tutsch, R.H. Earhart, *et al.* 1986. Phase I study of ICRF-187 in pediatric cancer patients and comparison of its pharmacokinetics in children and adults. *Cancer Treat Rep.* **70:** 703–709.
45. Mou, S.S., B.P. Giroir, E.A. Molitor-Kirsch, *et al.* 2004. Fresh whole blood versus reconstituted blood for pump priming in heart surgery in infants. *N. Engl. J. Med.* **351:** 1635–1644.

Ann. N.Y. Acad. Sci. ISSN 0077-8923

ANNALS OF THE NEW YORK ACADEMY OF SCIENCES
Issue: *Thymosins in Health and Disease*

Cardiac repair with thymosin β4 and cardiac reprogramming factors

Deepak Srivastava,[1,2,3] Masaki Ieda,[4,5] Jidong Fu,[1,2,3] and Li Qian[1,2,3]

[1]Gladstone Institute of Cardiovascular Disease, San Francisco, California. [2]Department of Pediatrics, University of California, San Francisco, California. [3]Department of Biochemistry and Biophysics, University of California, San Francisco, California. [4]Department of Clinical and Molecular Cardiovascular Research, Keio University School of Medicine, Tokyo, Japan. [5]Department of Cardiology, Keio University School of Medicine, Tokyo, Japan

Address for correspondence: Deepak Srivastava, M.D., Gladstone Institute of Cardiovascular Disease, 1650 Owens Street, San Francisco, CA 94158. dsrivastava@gladstone.ucsf.edu

Heart disease is a leading cause of death in newborns and in adults. We previously reported that the G-actin–sequestering peptide thymosin β4 promotes myocardial survival in hypoxia and promotes neoangiogenesis, resulting in cardiac repair after injury. More recently, we showed that reprogramming of cardiac fibroblasts to cardiomyocyte-like cells *in vivo* after coronary artery ligation using three cardiac transcription factors (Gata4/Mef2c/Tbx5) offers an alternative approach to regenerate heart muscle. We have combined the delivery of thymosin β4 and the cardiac reprogramming factors to further enhance the degree of cardiac repair and improvement in cardiac function after myocardial infarction. These findings suggest that thymosin β4 and cardiac reprogramming technology may synergistically limit damage to the heart and promote cardiac regeneration through the stimulation of endogenous cells within the heart.

Keywords: thymosin β4; cardiac reprogramming; myocardial infarction; cardiac repair

Introduction

Heart disease is the leading cause of death in the Western world.[1] Because the heart is incapable of sufficient muscle regeneration, survivors of myocardial infarctions typically develop chronic heart failure, with over five million cases in the United States alone.[1] Although more commonly affecting adults, heart disease in children is the leading noninfectious cause of death in the first year of life and often involves abnormalities in cardiac cell specification, migration, or survival.[2]

Recent evidence suggests that a population of extracardiac or intracardiac stem cells may contribute to maintenance of the cardiomyocyte population under normal circumstances.[3] Although the stem cell population may maintain a delicate balance between cell death and cell renewal, it is insufficient for myocardial repair after acute coronary occlusion. Introduction of isolated adult stem cells may improve myocardial function,[4–6] but this approach has been controversial, and may not contribute substantially to new muscle.[7,8] Pluripotent embryonic stem cells and induced pluripotent stem cells can be efficiently differentiated into the cardiomyocyte lineage using timed delivery of growth factors, and as a result, large numbers of cardiomyocytes can now be made from pluripotent stem cells. However, these cells remain immature, at least *in vitro*. Although technical hurdles of stem cell purification, delivery, and integration have thus far prevented clinical application of pluripotent stem cell–derived cardiomyocytes, the use of embryonic stem cell or induced pluripotent stem cell technology holds great promise for the future.

Regulatory pathways involved in cardiac development may have utility in cardiac repair.[9] In our studies of genes expressed during cardiac morphogenesis, we found that the forty-three amino acid peptide thymosin β4 was expressed in the developing heart. Thymosin β4 has numerous functions, with the most prominent involving sequestration of G-actin monomers and subsequent effects on actin-cytoskeletal organization necessary for cell motility,

doi: 10.1111/j.1749-6632.2012.06696.x

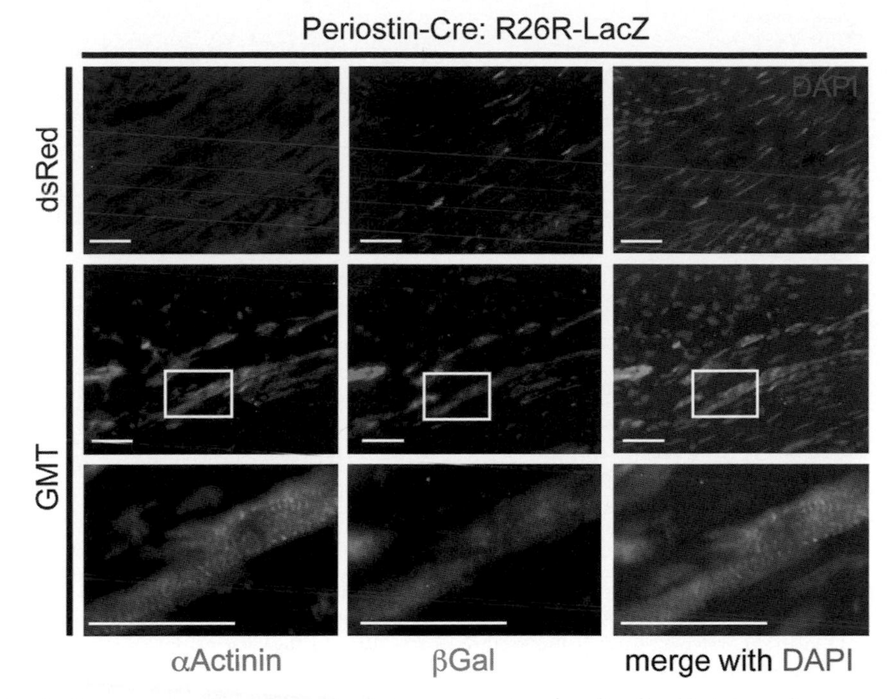

Figure 1. Genetic lineage tracing demonstrates *in vivo* reprogramming of cardiac fibroblasts to cardiomyocyte-like cells. Immunofluorescent staining for αActinin, βGal, and DAPI on injured areas of dsRed- or GMT-injected Periostin-Cre:R26R-lacZ mouse hearts four weeks post-MI. Boxed areas indicate regions of magnification shown in lower panels. Scale bar, 50 μm. Data are from Ref. 22.

organogenesis, and other cell biological events.[10–12] Domain analyses indicate that β-thymosins can affect actin assembly based on their carboxy-terminal affinity for actin.[13] Thymosin β4 can promote skin and corneal wound healing through its effects on cell migration, angiogenesis, and cell survival.[14–16] We previously reported that thymosin β4 can activate the survival kinase Akt and can play a potent role in protecting cardiac muscle from death after ischemic damage as occurs in the setting of a myocardial infarction.[17] Thymosin β4 can also promote angiogenesis in ischemic areas[18,19] and was reported to prime epicardial-derived progenitors to differentiate into cardiomyocytes.[20] Thus, there appear to be pleiotropic effects of thymosin β4 related to its role in promoting cardiac repair.

Most recently, we described the ability of three central cardiac developmental factors, Gata4, Tbx5, and Mef2c (GMT), to reprogram resident nonmyocytes in the heart into newly born cardiomyocyte-like cells.[21,22] We initially discovered this method *in vitro* and subsequently described its utility *in vivo* in the setting of myocardial infarction in a mouse model. Over half of all cells in the mammalian heart are cardiac fibroblasts that provide secreted signals to neighboring myocytes, and are activated after injury to form scar, but do not normally become muscle. Delivery of GMT into the nonmuscle population using retroviruses that can only infect dividing cells resulted in transdifferentiation of cardiac fibroblasts into cardiomyocyte-like cells that integrated with other myocytes *in vivo* and contributed to force generation. We also revealed that thymosin β4, in conjunction with delivery of GMT, could induce even greater cardiac repair than GMT alone in the setting of acute coronary occlusion in mice. We will describe the reprogramming process here and the effects of thymosin β4 in this new technology for cardiac repair.

Identification of cardiac reprogramming factors

To find a combination of reprogramming factors, we developed an assay to quantitatively analyze the reprogramming of fibroblasts toward the cardiomyocyte lineage by reporter-based fluorescence-activated cell sorting (FACS).[21] We generated

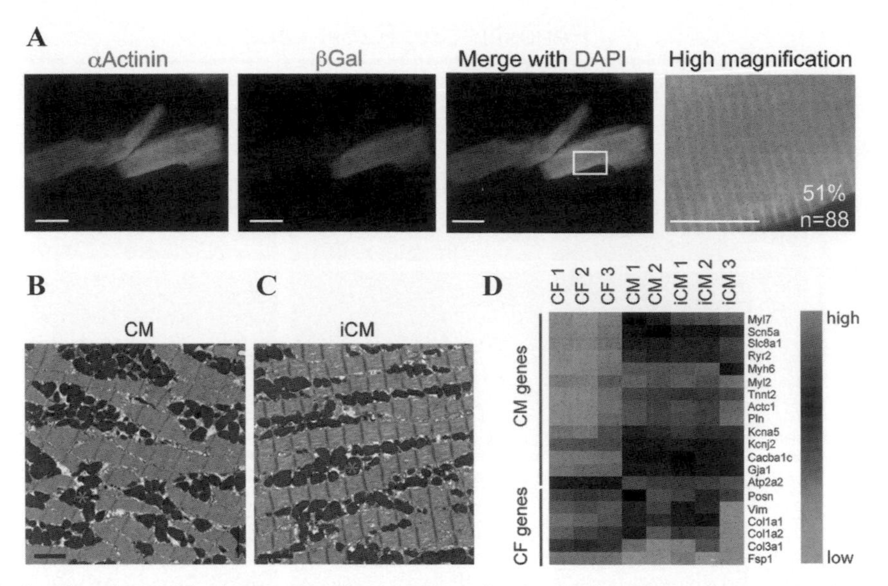

Figure 2. Single-cell analysis of the degree of cardiac reprogramming *in vivo.* (A) Immunofluorescent staining for αActinin, colabeled with βGal and DAPI, on isolated cardiomyocytes (CMs) from the infarct/border zone of periostin-Cre:R26R-lacZ hearts four weeks after GMT injection. Scale bar, 50 μm for the first three panels, 20 μm for the last panel. Red or yellow cells represent iCMs, green cells represent endogenous CMs. (B, C) Electron microscopy of endogenous CMs or iCMs. Asterisk indicates mitochondria and brackets indicate sarcomeric units. Scale bar, 2 μm. (D) Heat map of gene expression for a panel of CM- or fibroblast-enriched genes in isolated adult cardiac fibroblasts (CFs), CMs, or iCMs. Data are from Ref. 22.

αMHC promoter–driven transgenic mice (αMHC-green fluorescent protein), in which cardiomyocytes expressed GFP. Transduction of a mixture of retroviruses expressing 14 transcription factors involved in cardiac development into GFP-negative cardiac fibroblasts *in vitro* resulted in a small number of fibroblasts that became GFP$^+$ cells (1–2%), indicating activation of the cardiac-enriched αMHC gene. Serial deletion of each factor led to a combination of three transcription factors (Gata4, Mef2c, and Tbx5 (GMT)) that were necessary and sufficient to reprogram approximately 15% of postnatal cardiac fibroblasts into αMHC–GFP$^+$ cells that exhibited global reprogramming of gene expression.[21] The shift in cellular state was epigenetically stable and did not require ongoing ectopic gene expression of the reprogramming factors. Most cells assembled sarcomeres and displayed calcium transients, and ~30% expressed high levels of cardiac Troponin T (cTnT). However, less than 1% of *in vitro* reprogrammed cells spontaneously contracted at five weeks and had action potentials similar to adult ventricular cardiomyocytes, suggesting most were only partially reprogrammed. We termed the reprogrammed cells *induced cardiomyocytes* (iCMs). Others have found that fibroblasts can be directly converted to neuron-like cells,[23] hematopoietic-like cells,[24] and hepatocyte-like cells[25,26] using various cocktails of transcription factors and microRNAs.

In vivo cardiac repair with reprogramming technology

More recently, we introduced GMT directly into the mouse heart by intramyocardial injection of viral vectors to assay *in vivo* reprogramming of fibroblasts.[22] Using a genetic lineage-tracing approach, we showed that nonmyocytes in the heart, mostly cardiac fibroblasts, converted into cardiomyoctye-like cells four weeks after coronary ligation and intramyocardial injection of GMT (Fig. 1). Approximately one-third of cells in the border zone of injury that expressed sarcomeric genes were positive for β-Galactosidase (β-Gal) activity or the yellow fluourescent protein (YFP), activated by the fibroblast enriched promoter of Periostin or Fsp1 driving cre-recombinase. This result suggested the β-Gal$^+$ cells that developed sarcomeres were descendants of periostin or Fsp1-expressing nonmuscle cells. Importantly, we did not find evidence for cell fusion after genetically labeling endogenous cardiomyocytes and marking cells that were infected with retrovirus.

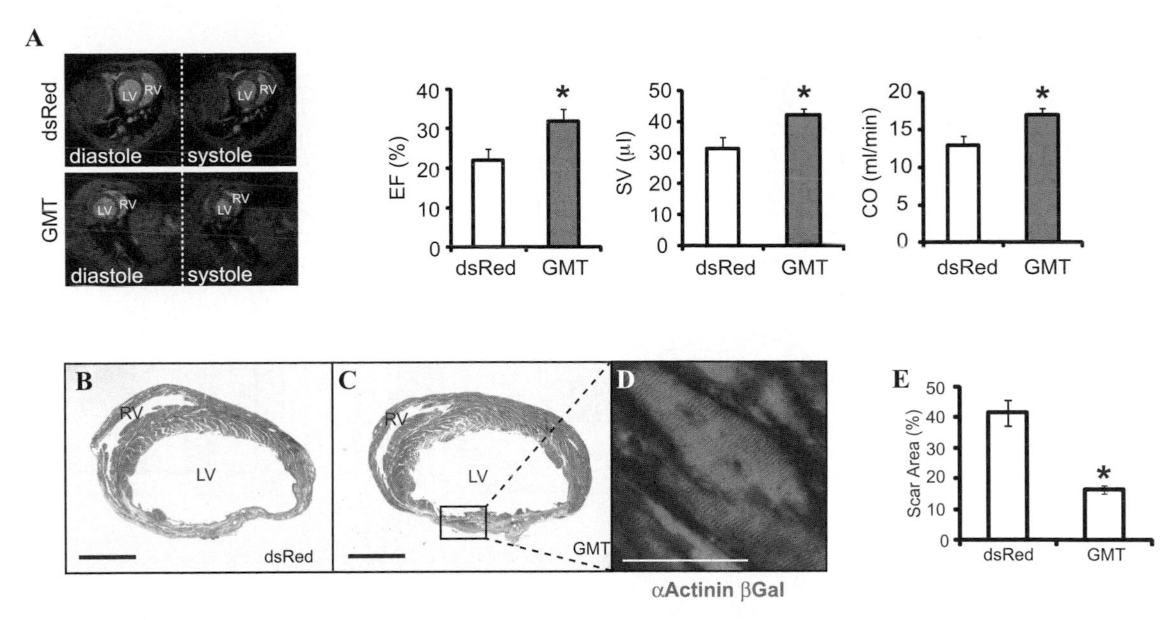

Figure 3. *In vivo* delivery of cardiac reprogramming factors improves cardiac function after myocardial infarction. (A) Ejection fraction (EF), stroke volume (SV), and cardiac output (CO) were quantified by MRI (left panels) 12 weeks after MI ($n = 9$/group, $^*P < 0.05$). Representative MRI thoracic images are shown in diastole and systole with dsRed or GMT injection. (B, C) Masson–Trichrome staining on heart sections of Periostin-Cre: R26RLacZ mice eight weeks post-MI injected with dsRed or GMT. Scale bars, 500 μm. (D) α Actinin + βGal$^+$ cells in the infarct area. Scale bar, 50 μm. (E) Calculation of scar area (dsRed, $n = 8$; GMT, $n = 9$; $^*P < 0.05$). Scale bars: 500 μm. Error bars indicate standard error of the mean (S.E.M.). Data are from Ref. 22.

Morphologically, half of iCMs were large with a rod-shaped appearance and were binucleated, closely resembling endogenous cardiomyocytes from the same isolation. Further analyses revealed that, in addition to α-actinin, the β-galactosidase$^+$ cells expressed multiple sarcomeric markers, including tropomyosin, cardiac muscle heavy chain (αMHC), and cTnT. Examples of cells that showed nearly normal sarcomeric structures throughout the cell, representing ~50% of cells, are shown in Figure 2A, while others had intermediate levels of sarcomeric organization in the cell. Similar results were obtained by electron microscopy, with a range of ultrastructural organization, although almost all had well-developed and abundant mitochondria (Fig. 2B and C). qPCR on a panel of 20 genes normally enriched in CMs or cardiac fibroblasts indicated the gene expression of iCMs was similar to that of CMs (Fig. 2D). Importantly, isolated cells also expressed and localized gap junction proteins, such as connexin 43 and N-cadherin. Small dyes could pass through the gap junction, and calcium waves could be propagated from an induced cardiomyocyte to an endogenous cardiomyocyte. Individual YFP$^+$ iCMs isolated from the heart had ventricular-like action potentials by patch-clamp studies and ~50% of isolated YFP$^+$ cells beat similarly to endogenous cardiomyocytes upon stimulation, while others were incompletely depolarized, suggestive of only partial reprogramming.

Because *in vivo* reprogrammed iCMs had contractile potential and electrically coupled with viable endogenous cardiomyocytes, we determined if converting endogenous cardiac fibroblasts into myocytes translates into partial restoration of heart function after MI. Mice injected with GMT or dsRed alone into the border/infarct zone immediately after coronary ligation were evaluated after three months for cardiac function by magnetic resonance imaging (MRI) and high resolution 2-D echocardiography. MRI provides the most accurate three-dimensional assessment of the fraction of blood ejected with each ventricular contraction (ejection fraction), the volume of blood ejected (stroke volume), and the total cardiac output per minute (Fig. 3A). Each of these parameters was significantly improved with GMT infection, particularly the stroke volume and cardiac output, possibly due to cardiac enlargement. Serial echocardiography showed similar results, with improvement detectable by eight weeks after injection.

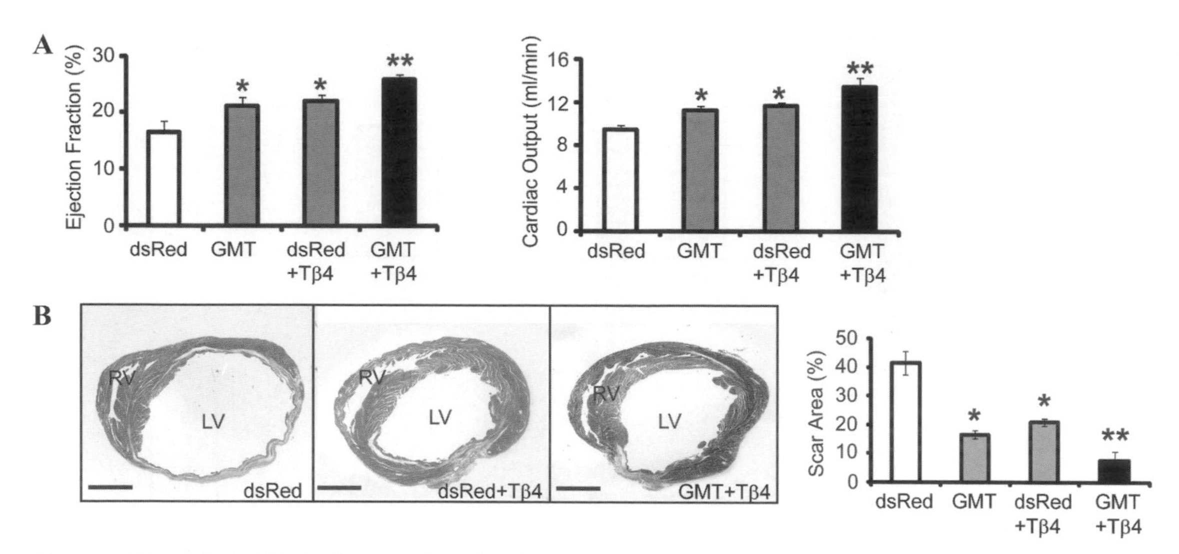

Figure 4. Thymosin β4 (Tβ4) enhances the benefits of *in vivo* cardiac reprogramming with GMT. (A) Ejection fraction (EF) and cardiac output (CO) of the left ventricle were determined using high-resolution echocardiography eight weeks postsurgery after injection of dsRed ($n = 9$); GMT ($n = 10$); dsRed + Tβ4 ($n = 10$); GMT + Tβ4 ($n = 8$); ($^*P < 0.05$, $^{**}P < 0.01$). (B) Scar area was calculated in a blinded fashion from multiple heart sections eight weeks post-MI after dsRed ($n = 8$), GMT ($n = 9$), dsRed + Tβ4 ($n = 7$), or GMT + Tβ4 ($n = 8$) injection. Representative Masson–Trichrome staining on heart sections is shown. Scale bars: 500 μm. Quantification of scar size was calculated by measuring the scar area in a blinded fashion. $^*P < 0.05$, $^{**}P < 0.01$. Error bars indicate standard error of the mean (S.E.M.). Data are from Ref. 17.

In agreement with the improvement of cardiac function, calculation of scar area with 16 sections at four levels of the heart revealed a significantly smaller scar size eight weeks after MI in the GMT-treated group, with presence of iCMs in the area of scar (Fig. 3B–E).

Enhanced cardiac repair with thymosin β4 and cardiac reprogramming factors

We and others had previously shown that thymosin β4 could promote cardiac cell migration, activate proliferation of cardiac fibroblasts and endothelial cells, and promote neoangiogenesis.[17–19] One mechanism by which thymosin β4 functions is through interaction with PINCH and integrin linked kinase (ILK) as part of a larger complex involved in cell-extracellular matrix interactions known as the focal adhesion complex. PINCH and ILK are required for cell motility[27,28] and for cell survival, in part, by promoting phosphorylation of the serine-threonine kinase Akt/protein kinase B, a central kinase in survival and growth signaling pathways.[27–30] All three proteins could be isolated as a complex, and we have demonstrated that thymosin β4 induces activation of signaling events downstream of ILK, particularly, phosphorylation and activation of Akt.[17]

In rodent and pig models of myocardial infarction, thymosin β4 had potent effects in limiting the amount of damage caused by coronary ligation.[17,31] Thymosin β4 administration appeared to promote cell survival and initiate neoangiogenesis in the area of hypoxia. More recently, thymosin β4 administration before injury appeared to prime a population of epicardium-derived progenitor cells to become new cardiomyocytes.[20]

Although GMT delivery significantly affected cardiac repair after MI, we hypothesized that concurrent administration of thymosin β4 might further enhance the degree of repair, in part by increasing the number of fibroblasts that became activated and proliferative, and through promoting angiogenesis. To test the effects of thymosin β4 on cardiac fibroblast migration, we used a cardiac explant migration assay.[17,21] The average time for fibroblasts to migrate from adult heart explants was approximately three weeks; however, with thymosin β4, equivalent fibroblast migration was observed after only two weeks and occurred within three days after MI. Similarly, the proliferation of Vimentin$^+$ cells increased after MI and was even more increased following administration of thymosin β4, as marked by phosphohistone H3. Consistent with the activation of fibroblasts by thymosin β4, the percent of Thy1$^+$ or vimentin$^+$ cells (markers of fibroblasts) infected by retrovirus in the setting of MI more than doubled upon intramyocardial injection of thymosin β4. The improved delivery of GMT-expressing retrovirus to more cells by addition of thymosin β4 resulted in an increase in the percentage of iCMs compared to total CMs in single-cell CM culture from the infarct/border zone of periostin-Cre:R26R-lacZ hearts (51% vs. 35%, $P < 0.05$).

Injection of thymosin β4 immediately after coronary ligation resulted in improvement of cardiac function, as previously reported.[17,31] Coinjection of thymosin β4 and GMT yielded further functional improvement in ejection fraction and cardiac output eight weeks after infarction (Fig. 4A). In agreement with this, coinjection of thymosin β4 and GMT caused a greater reduction in scar size than either thymosin β4 or GMT injection alone (Fig. 4B), despite the area at risk and initial infarct size being similar in both groups. The combined effects of GMT and thymosin β4 on cardiac function were particularly powerful and suggest that the two approaches may synergize with one another to limit damage to the heart and promote cardiac repair.

The full mechanism of cardiac repair with GMT and/or with thymosin β4 remains to be determined. Future studies examining the DNA-occupancy of the critical transcription factors will be informative, as will evaluation of the time course of the reprogramming process. It is possible that the small population of epicardium-derived cells that may give rise to new myocytes after pretreatment with thymosin β4 could be contributing to the effects observed in our study, although no effect was observed by Smart *et al.* when thymosin β4 was given at the time of injury.[20] Nevertheless, it is possible that transduction of GMT into these progenitors, or other rare progenitors yet to be identified, could promote their differentiation into cardiomyocytes. Because thymosin β4 is also a proangiogenic factor,[18,19] the full mechanism of cooperativity between GMT and thymosin β4 may be multifaceted and will be interesting to explore.

Acknowledgments

D.S. was supported by grants from NHLBI/NIH, the California Institute of Regenerative Medicine (CIRM), the Roddenberry Foundation, the Younger Family Foundation, and the L.K. Whittier Foundation. J.F. was supported by a postdoctoral fellowship from the American Heart Association. M.I. was supported by research grants from JST CREST and JSPS. L.Q. was supported by a CIRM postdoctoral fellowship.

Conflicts of interest

The authors declare no conflicts of interest.

References

1. Roger, V.L., A.S. Go, D.M. Lloyd-Jones, *et al.* 2012. Heart disease and stroke statistics–2012 update: a report from the American Heart Association. *Circulation* **125:** e2–e220.
2. Hoffman, J.I. & S. Kaplan. 2002. The incidence of congenital heart disease. *J. Am. Coll. Cardiol.* **39:** 1890–1900.
3. Bergmann, O., R.D. Bhardwaj, S. Bernard, *et al.* 2009. Evidence for cardiomyocyte renewal in humans. *Science* **324:** 98–102.
4. Orlic, D., J. Kajstura, S. Chimenti, *et al.* 2001. Bone marrow cells regenerate infarcted myocardium. *Nature* **410:** 701–705.
5. Beltrami, A.P., L. Barlucchi, D. Torella, *et al.* 2003. Adult cardiac stem cells are multipotent and support myocardial regeneration. *Cell* **114:** 763–776.
6. Anversa, P. & B. Nadal-Ginard. 2002. Myocyte renewal and ventricular remodelling. *Nature* **415:** 240–243.
7. Balsam, L.B., A.J. Wagers, J.L. Christensen, *et al.* 2004. Haematopoietic stem cells adopt mature haematopoietic fates in ischaemic myocardium. *Nature* **428:** 668–673.
8. Murry, C.E., M.H. Soonpaa, H. Reinecke, *et al.* 2004. Haematopoietic stem cells do not transdifferentiate into cardiac myocytes in myocardial infarcts. *Nature* **428:** 664–668.
9. Srivastava, D. 2006. Making or breaking the heart: from lineage determination to morphogenesis. *Cell* **126:** 1037–1048.
10. Safer, D., M. Elzinga & V. Nachmias. 1991. Thymosin beta 4 and Fx, an actin-sequestering peptide, are indistinguishable. *J. Biol. Chem.* **266:** 4029–4032.
11. Huff, T., C.S. Muller, A.M. Otto, *et al.* 2001. Beta-Thymosins, small acidic peptides with multiple functions. *Int. J. Biochem. Cell Biol.* **33:** 205–220.
12. Sun, H., K. Kwiatkowska & H. Yin. 1996. Beta-Thymosins are not simple actin monomer buffering proteins. Insights from overexpression studies. *J. Biol. Chem.* **271:** 9223–9230.
13. Hertzog, M., C. van Heijenoort, D. Didry, *et al.* 2004. The beta-thymosin/WH2 domain; structural basis for the switch from inhibition to promotion of actin assembly. *Cell* **117:** 611–623.
14. Malinda, K., G. Sidhu, H. Mani, *et al.* 1999. Thymosin beta4 accelerates wound healing. *J. Invest. Dermatol.* **113:** 364–368.
15. Sosne, G., E. Szliter, R. Barrett, *et al.* 2002. Thymosin beta 4 promotes corneal wound healing and decreases inflammation *in vivo* following alkali injury. *Exp. Eye. Res.* **74:** 293–299.
16. Grant, D.S., W. Rose, C. Yaen, *et al.* 1999. Thymosin beta 4 enhances endothelial cell differentiation and angiogenesis. *Angiogenesis* **3:** 125–135.
17. Bock-Marquette, I., A. Saxena, M.D. White, *et al.* 2004. Thymosin beta4 activates integrin-linked kinase and promotes cardiac cell migration, survival and cardiac repair. *Nature* **432:** 466–472.

18. Smart, N., C.A. Risebro, A.A. Melville, *et al.* 2007. Thymosin beta4 induces adult epicardial progenitor mobilization and neovascularization. *Nature* **445:** 177–182.
19. Bock-Marquette, I., S. Shrivastava, G.C. Pipes, *et al.* 2009. Thymosin beta4 mediated PKC activation is essential to initiate the embryonic coronary developmental program and epicardial progenitor cell activation in adult mice *in vivo*. *J. Mol. Cell. Cardiol.* **46:** 728–738.
20. Smart, N., S. Bollini, K.N. Dube, *et al.* 2011. De novo cardiomyocytes from within the activated adult heart after injury. *Nature* **474:** 640–644.
21. Ieda, M., J. Fu, P. Delgado-Olguin, *et al.* 2010. Direct reprogramming of fibroblasts into functional cardiomyocytes by defined factors. *Cell* **142:** 375–386.
22. Qian, L., Y. Huang, C.I. Spencer, *et al.* 2012. *In vivo* reprogramming of murine cardiac fibroblasts into induced cardiomyocytes. *Nature* **485:** 593–598.
23. Vierbuchen, T., A. Ostermeier, Z.P. Pang, *et al.* 2010. Direct conversion of fibroblasts to functional neurons by defined factors. *Nature* **463:** 1035–1041.
24. Szabo, E., S. Rampalli, R.M. Risueno, *et al.* 2010. Direct conversion of human fibroblasts to multilineage blood progenitors. *Nature* **468:** 521–526.
25. Huang, P., Z. He, S. Ji, *et al.* 2011. Induction of functional hepatocyte-like cells from mouse fibroblasts by defined factors. *Nature* **475:** 386–389.
26. Sekiya, S. & A. Suzuki. 2011. Direct conversion of mouse fibroblasts to hepatocyte-like cells by defined factors. *Nature* **475:** 390–393.
27. Fukuda, T., K. Chen, X. Shi & C. Wu. 2003. PINCH-1 is an obligate partner of integrin-linked kinase (ILK) functioning in cell shape modulation, motility, and survival. *J. Biol. Chem.* **278:** 51324–51333.
28. Zhang, Y., K. Chen, Y. Tu, *et al.* 2002. Assembly of the PINCH-ILK-CH-ILKBP complex precedes and is essential for localization of each component to cell-matrix adhesion sites. *J. Cell. Sci.* **115:** 4777–4786.
29. Troussard, A., N. Mawji, C. Ong, *et al.* 2003. Conditional knock-out of integrin-linked kinase demonstrates an essential role in protein kinase B/Akt activation. *J. Biol. Chem.* **278:** 22374–2278.
30. Brazil, D.P., J. Park & B.A. Hemmings. 2002. PKB binding proteins. Getting in on the Akt. *Cell* **111:** 293–303.
31. Hinkel, R., C. El-Aouni, T. Olson, *et al.* 2008. Thymosin beta4 is an essential paracrine factor of embryonic endothelial progenitor cell-mediated cardioprotection. *Circulation* **117:** 2232–2240.

Ann. N.Y. Acad. Sci. ISSN 0077-8923

ANNALS OF THE NEW YORK ACADEMY OF SCIENCES
Issue: *Thymosins in Health and Disease*

NMR structural studies of thymosin α1 and β-thymosins

David E. Volk,[1,2] Cynthia W. Tuthill,[3] Miguel-Angel Elizondo-Riojas,[1,4] and David G. Gorenstein[1,2]

[1]Institute of Molecular Medicine, [2]Department of Nanomedicine and Biomedical Engineering, University of Texas Health Science Center, Houston, Texas. [3]SciClone Pharmaceuticals, Inc., Foster City, California. [4]Centro Universitario Contra el Cáncer, Hospital Universitario "Dr. José Eleuterio González," Universidad Autónoma de Nuevo León, Monterrey, Mexico

Address for correspondence: David E. Volk, Institute of Molecular Medicine, University of Texas Health Science Center, 1825 Pressler, Houston, TX 77030. David.Volk@uth.tmc.edu

Thymosin proteins, originally isolated from fractionation of thymus tissue, represent a class of compounds that we now know are present in numerous other tissues, are unrelated to each other in a genetic sense, and appear to have different functions within the cell. Thymosin α1 (generic drug name thymalfasin; trade name Zadaxin) is derived from a precursor molecule, prothymosin, by proteolytic cleavage, and stimulates the immune system. Although the peptide is natively unstructured in aqueous solution, the helical structure has been observed in the presence of trifluoroethanol or unilamellar vesicles, and these studies are consistent with the presence of a dynamic helical structure whose sides are not completely hydrophilic or hydrophobic. This helical structure may occur in circulation when the peptide comes into contact with membranes. In this report, we discuss the current knowledge of the thymosin α1 structure and similar properties of thymosin β4 and thymosin β9, in different environments.

Keywords: thymosin α1; thymosin β4; thymosin β9; thymalfasin; protein structure; immunogenic peptide hormone

Introduction

Thymosin α1 is a 28 amino acid immunogenic thymic peptide, with an N-terminal acetate cap, first isolated from bovine thymus and characterized by Allan Goldstein in 1977.[1] It is being used as an adjuvant for the enhancement of immunity; recent studies have investigated the use of thymosin α1 and its precursor, prothymosin, in the treatment of chronic hepatitis B,[2,3] hepatitis C,[4,5] HIV,[6,7] cytomegalovirus,[8] invasive aspergillosis,[9] and dendritic cell tryptophan catabolism.[10] Thymosin α1 has been approved in many countries for vaccine enhancement and for the treatment of chronic hepatitis B and C as well as certain cancers.

Structures of thymosin proteins in water

Thymosin α1[11] and the beta-thymosin peptides, thymosin β4[12] and thymosin β9,[13] are all known to be natively unstructured in water solutions and ambient temperatures, but helical and beta segments can be detected in them at lower temperatures or in the presence of fluoroalcohols such as 1,1,1,3,3,3-hexafluoro-2-propanol (HFP) or trifluoroethanol (TFE).[11–14]

Structure of thymosin β4 in water

Czisch *et al.*[12] reported NMR-based structural information for the 43-amino acid peptide thymosin β4 by investigating the effects of salt concentration, temperature and pH. They investigated thymosin β4 structure at pH = 3, at temperatures of 1, 4, 14, and 29 °C, and at pH = 6.5 at temperatures 1 and 14 °C. They found that salt concentration had a negligible effect on the NMR spectra, but the temperature played a large role in the structure. At both 1 and 4 °C, they observed strong HN-HN NOE signals indicative of a helical structure extending between residues 5–19 and 30–37. However, at 14 °C, both the HN-HN sequential NOEs and other medium range NOE crosspeaks disappeared for the C-terminal helix, indicating that the helical structure for residues 30–37 had been melted away. They also estimated, based on circular dichroism (CD),

doi: 10.1111/j.1749-6632.2012.06656.x

that thymosin β4 in water contains 15% helix at 4 °C, 9% helix at 9 °C, and 7% helix at 20 °C.

Structure of thymosin β9 in water

Subsequently, Stoll *et al.*[13] investigated the structure of the 41 amino acid peptide thymosin β9, first isolated from calf thymus tissue,[15] in water and in a 40% HFP solution. Both CD and NMR measurements indicated that in water thymosin β9 lacks secondary structure, but that secondary structural elements are present in 40% HFP solution.

Structure of thymosin α1 in water

Grottesi *et al.*[11] noted that thymosin α1 also does not form a preferred structure, but that secondary structural elements are present in a 40% TFE solution, as further elucidated by Elizondo-Riojas *et al.*[14] (see below).

Structure of thymosin α1 in membrane-like environments

Grottessi *et al.*[11] also investigated the structure of thymosin α1 in the presence of unilamellar phosphatidylcholine vesicles or sodium dodecyl sulphate (SDS), which are both negatively charged membrane-like environments, by both CD and NMR measurements to model its possible interactions with a lymphocyte membrane. They first investigated the thymosin α1 structure in the presence of dimyristoylphosphatidylcholine (DMPC) with thymosin α1 at molar ratios of 3:1, 1:1, and 1:3, and found no structural changes by CD relative to thymosin α1 in water. However, notable structural changes were observed for thymosin α1 by CD spectra in solutions containing both DMPC and dimyristoylphosphatidic acid (DMPA) at DMPC:DMPA ratios of either 10:1 or 9:1 in the presence of zinc ions. These changes corresponded to increased helical structure in the peptide. They reported similar results in the presence of SDS.

NMR structure of thymosin α1 in 40% TFE/60% H_2O

Overall, the NMR structure of thymosin α1 in 40% TFE can be described as two relatively structured regions connected by a flexible hinge. As shown in Figure 1A, the N-terminal amino acids Ser1 to Ile11 for the ensemble of NMR structures reported by Elizondo-Riojas *et al.*[14] align very tightly when fit to these residues, but the remaining amino acids

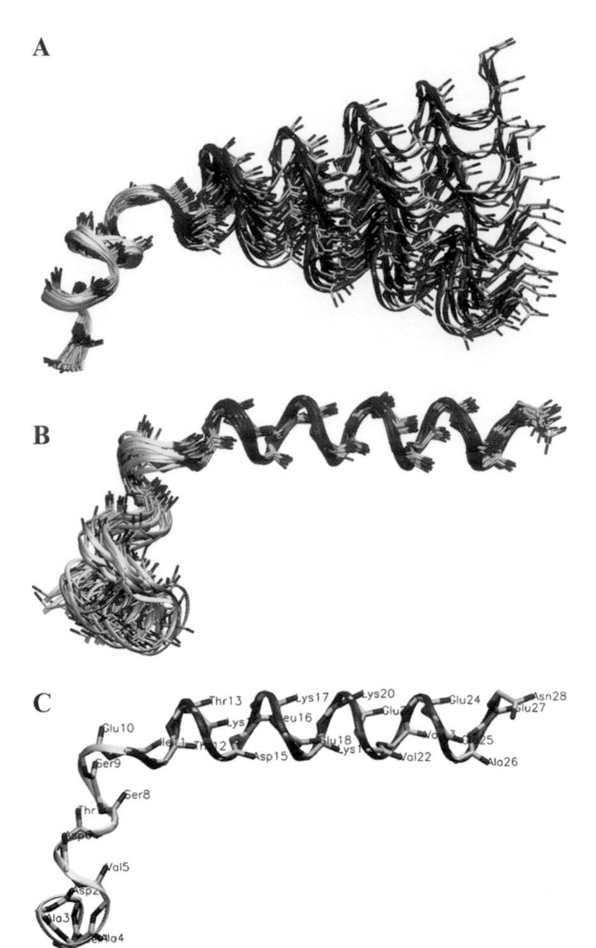

Figure 1. From the 20 lowest energy structures of thymosin α1 in 40% TFE/60% water and the average structure.[14] (A) Ribbon representation of the backbone superposition of the N-terminal region fitted to residues 1–11 (yellow). (B) Ribbon representation of the backbone superposition of the C-terminal region fitted to residues 12–28 (violet). (C) Ribbon representation of the backbone of the average structure (yellow: residues 1–11; violet: residues 12–28).

Thr12 to Asn28 align poorly. Likewise, when the protein structure ensemble is aligned for amino acids Thr12 to Asn28 (Fig. 1B), the C-terminal helix from Thr12 to Asn28 aligns tightly, leaving the first eleven residues poorly aligned.

The C-terminal helix of thymosin α1 in 40% TFE/60% H_2O

The presence of a helical structure in the C-terminal portion of thymosin α1 was first noted by Grottessi *et al.*[11] They described the presence of a helix from residues K14 to E24 when in the presence of 40%

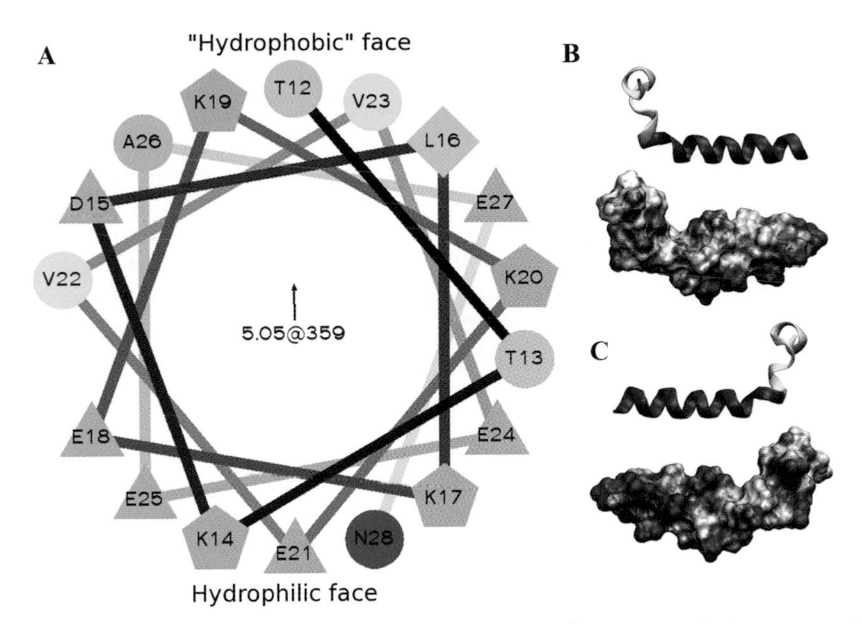

Figure 2. From the average structure of thymosin α1 in 40% TFE/60% water.[14] (A) Helical wheel projection of residues 12–28. (B,C) Electrostatic surface potential of the "hydrophobic" and hydrophilic faces of thymosin α1, respectively (basic: blue; acidic: red).

TFE. However, this description was based on nuclear magnetic resonance data acquired at only 400 MHz, and, thus, that study suffered from both a relative lack of resolution and signal compared to the later study of Elizondo-Riojas *et al.*,[14] which was conducted at 800 MHz. Owing to the lower signal intensity and, more importantly, lower resolution, the earlier work was unable to complete all of the proton assignments for residues E25, E27, and N28. As a consequence of this, the NMR structure determination lacked any NOE-derived restraints with which to indicate the presence of a helical structure in this area. The Grottesi structure also suffered from apparent incorrect resonance assignments for the alpha protons of residues on the N-terminal side of the C-terminal helix, namely the alpha protons of residues I11, T12, T13, and L16.

The C-terminal helix is made up of residues T12 to N28 (Fig. 1C), and a complete network of hydrogen bonds between backbone carbonyl oxygen atoms O(i) and amide hydrogen atoms Hn(i+4) is observed in the structure. It is interesting to note that residues S8, S9, E10, and I11 exhibit what appears to be a distorted helix that is distended away from the C-terminal helix. However, these residues are part of a double beta-turn, as discussed below. The C-terminal helix observed in 40% TFE is not a classical amphiphilic helix (Fig. 2A). One face of the helix, composed of residues E18, E25, K14, E21, N28, K17, E24, T13, and K20 is very hydrophilic, containing seven charged hydrophilic residues, and two polar hydrophilic residues, Thr13 and Asn28. The "hydrophobic" side of the helix contains hydrophobic residues V22, A26, V23, and L16, but it also contains hydrophilic charged amino acids D15, K19, E27, polar hydrophilic residue T12, and possibly also K20, depending on how one defines the sides. The helix also contains an asymmetric charge distribution along the helical axis, as is often seen in helices. All four lysine residues in the helix, Lys14, Lys17, Lys19, and Lys20, occur in the N-terminal half of the helix, while the C-terminal half of the helix contains four (Glu21, Glu24, Glu25, Glu27) of the five acidic groups. Only Asp15 is present in the N-terminal half of the helix.

This can be seen more clearly in the electrostatic surface potentials (Figs. 2B and C). For the hydrophilic side (Fig. 2C), the positively charged side chains of K14 and K17 are observed as blue surfaces on the bottom, while the acidic residues E18, E21, E24, and E25 are depicted in red. Very little hydrophobic surface is indicated by white coloring. In contrast, the "hydrophobic" side (Fig. 2B) is clearly a mix of charged and neutral species, as indicated by mostly white neutral surface area with ample charged surface groups mixed in.

The N-terminal structure of thymosin α1 in 40% TFE/60% H_2O by NMR

Grottesi *et al.*[11] first reported the presence of beta structure in the N-terminal portion of thymosin α1, based in part to the observation of strong NOE signals of the type αN(i, i+1) between A4-V5, and between V5 and D6, which is indicative of a beta-turn. In that work, the beta structure is described as occurring between residues V5 to S8. The subsequent structure reported by Elizondo-Riojas *et al.*[14] determined a structure for the N-terminal residues that, if not for a strange kink between residues D6 and T7, could be loosely described as helix pulled apart. They also noted large αN(i, i+1) NOE signals for residues A4 to V5 and V5 to D6. However, a more exacting description of the later structure's N-terminal region is that it contains two double beta-turns type I (Fig. 3), as defined by Hutchinson and Thornton.[16]

The first double beta-turn is an (i, i+1)-type of double beta-turn made up of two sequential type-I beta turns. The first type-I beta turn occurs at residues D2 to V5, as shown in Figure 3(A). The distance between the Asp2 carbonyl oxygen atom and the amide nitrogen of Val5 is 3.0 Å, indicative of a strong hydrogen bond. The second type-I beta-turn within the first double beta-turn occurs along residues Ala3 to Asp6, as shown in Fig. 3B. The carbonyl oxygen of Ala3 is 3.4 Å from the amide nitrogen of Asp6, indicating a slightly weaker hydrogen bond.

The second double beta-turn is an (i, i+2)-type of double beta-turn made up of nonsequential type-I beta turns. The first type-I beta-turn occurs along residues Thr7 to Glu10, as shown in Fig. 3C. The Thr7 carbonyl oxygen is 3.3 Å from the Glu10 amide nitrogen atom. The second type-I beta-turn within this double beta-turn is formed by residues Ser9 to Thr12 (Fig. 3D). This is a rather wide beta turn, perhaps because Thr12 is the N-terminal end of the C-terminal alpha helix, and due to the relative flexibility of the connection between residues Ile11 and Thr12 (Fig. 1).

NMR structure of thymosin β4 in 50% HFP/50% H_2O or in 60% TFE/40% H_2O

Zarbock *et al.*[17] investigated the conformational changes that occur to thymosin β4 in the presence of mixed solutions of water (H_2O or D_2O) and one of two fluoroalcohols, HFP or TFE. In the fluoroalcohols, the line widths of the protein's resonances were significantly increased (compared to in water), indicating slower rotational movement or slower interconversion between conformers and likely formation of some secondary structure. The two fluoroalcohols give essentially identical NOE

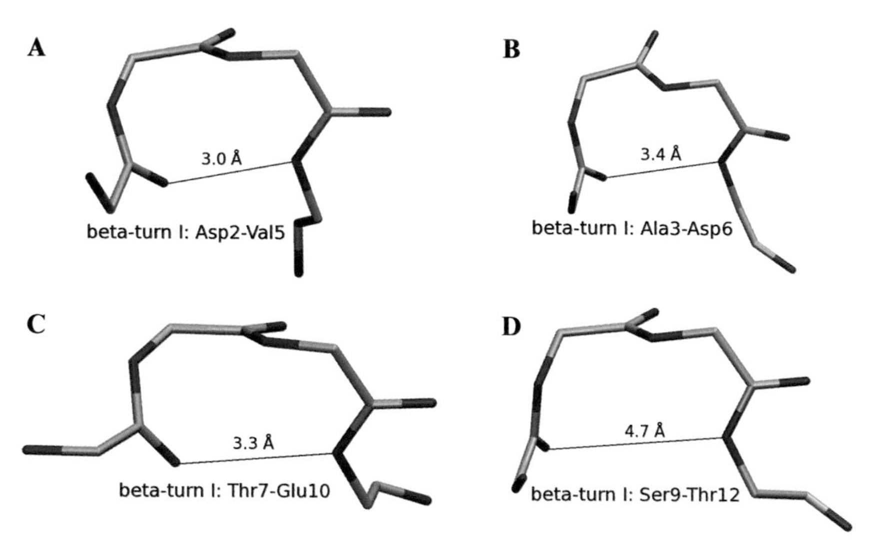

Figure 3. Licorice representation of the two double beta-turns type I from the N-terminal region of the average structure of thymosin α1 in 40% TFE/60% water.[14] (A,B) First double beta-turn (i, i+1)-type formed by Asp2-Ala3-Ala4-Val5 and Ala3-Ala4-Val5-Asp6. (C,D) Second double beta-turn (i, i+2)-type formed by Thr7-Ser8-Ser9-Glu10 and Ser9-Glu10-Ile11-Thr12.

patterns, but slight differences in the chemical shifts between solvents helped in the resonance assignment step. Under these conditions, nearly complete proton NMR assignments were made for this 43-amino acid peptide, although few assignments were made for Lys18 and Lys19 and several other resonance frequencies remained unassigned. With sequential HN(i)-HN(i+1) correlations and medium range αN(i, i+3), αN(i, i+4), and αβ(i, i+3) correlations, they found the presence of two helices extending from residues Pro4 (or Asp5) to Lys16 and from Ser30 to Ala40. The first helix has an asymmetrical preference for negative amino acids on the N-terminal end and for positive amino acids on the C-terminal end. This was later also noted for thymosin β9, as noted below. The proline amino acid at position N1 of the first helix is consistent with preferences noted by Richardson & Richardson,[18] and the Gly41 N-cap, and the Ser30 C-cap of the second helix, fit known preferences for helix caps.[19–22] Because of the lack of long-range NOE interactions between residues more than four residues apart, the relative orientation of the two helices with respect to one another could not be determined.

NMR structure of thymosin β9 in 40% HFP/60% H_2O

Although thymosin β9 is natively unstructured in water, in a solution of 40% HFP, Stoll *et al.*[13] found that two helices were formed that encompassed nearly the entire peptide. Under these conditions, essentially complete proton NMR assignments were accomplished, and strong sequential HN-HN NOE cross peaks and medium range αN(i, i+3) and αN(i, i+4), all indicative of a helical structure, were observed.

The first helix, spanning residues Pro4 to Thr27, and the second helix, spanning residues Thr32 to Lys41, are connected by a flexible loop region spanning residues Leu28 to Lys31. The first helix shows a preference for negatively charges amino acids on the N-terminal end and for positively charged amino acids on the C-terminal end. A similar arrangement was previously noted in thymosin β4, while the main helix in thymosin α1 has the opposite arrangement, containing more positively charged amino acids toward the N-terminal end and all acidic amino acids on the C-terminal end of the helix.

NMR structure of thymosin β4 when interacting with monomeric CaATP-actin

Thymosin β4 has been identified as the most abundant actin-sequestering protein in platelets[23,24] and neutrophils.[25] It contains a single WASP homology 2 (WH2) domain, an evolutionarily conserved actin-monomer–binding motif found in many proteins.[26] Determining the structure of thymosin β4 bound to G-actin has proven difficult, due to the propensity of G-actin to polymerize to F-actin at high concentrations, but X-ray structures of actin bound to DNase 1,[27] gelsolin segment-1,[28] and profilin[29] have been reported. Domanski *et al.*[30] reported the NMR structure of thymosin β4 upon interaction with monomeric actin using uniformly and selectively ^{15}N-labeled proteins. At 2 °C, before addition of the actin, they observed NOE signals, indicating a partial helical structure along residues Asp5 to Lys16 (in exchange with an extended beta structure), in agreement with the earlier report by Czisch *et al.*,[12] and chemical shifts differing from random coil, suggesting a *weak tendency* for residues Lys31 to Glu37 to form a helix, in agreement with the helix observed in TFE by Zarbock *et al.*[17] After addition of actin, a large protein by NMR standards, Domanski *et al.*[30] observed an extended central fragment, from Lys18 to Asn26, and one alpha helix on either end, at residues Asp5 to Leu17 and Lys31 to Ala40. The HSQC spectra of thymosin β10 exhibited a similar dispersion of signals upon mixing with actin.

Conclusion

While thymosin α1 has been shown to be natively unstructured in an aqueous environment at ambient temperatures, it forms a helical structure in the presence of TFE or unilamellar vesicles, and these studies are consistent with the presence of a dynamic helical amphipathic structure whose sides are not completely hydrophilic or hydrophobic. This helical structure may occur in circulation when the peptide comes into contact with membranes, which could be involved in assisting thymosin α1 to cross the membrane, consistent with the fact that thymosin α1 interacts with Toll-like receptor 9,[8] an intracellular molecule. Likewise, the beta thymosins β4 and β9 are inherently unstructured at ambient temperatures in an aqueous environment, but the presence of fluoroalcohols or a binding partner such as actin induces the formation of helices.

Conflicts of interest

C.W. Tuthill works for SciClone Pharmaceuticals, Inc.

References

1. Goldstein, A.L, T.L. Low, M. McAdoo, *et al.* 1977. Thymosin α1: isolation and sequence analysis of an immunologically active thymosin polypeptide. *Proc. Natl. Acad. Sci. USA* **74:** 725–729.
2. Iino, S., J. Toyota, H. Kumada, *et al.* 2005. The efficacy and safety of thymosin alpha-1 in Japanese patients with chronic hepatitis B; results from a randomized clinical trial. *J. Viral Hepat.* **12:** 300–306.
3. You, J., L. Zhuang, H.-Y. Cheng, *et al.* 2006. Efficacy of thymosin alpha-1 and interferon alpha in treatment of chronic viral hepatitis B: a randomized controlled study. *World J. Gastroenterol.* **12:** 715–721.
4. Andreone, P., C. Cursaro, A. Gramenzi, *et al.* In vitro effect of thymosin-α1 and interferon-α on Th1 and Th2 cytokine synthesis in patients with chronic hepatitis C. *J. Viral Hepatitis* **8:** 194–201.
5. Kullavanuaya, P., S. Treeprasertsuk, D. Thong-Ngam, *et al.* 2001. The combined treatment of interferon alpha-2a and thymosin alpha 1 for chronic hepatitis C: the 48 weeks end of treatment results. *J. Med. Assoc. Thai.* **84:** 462–468.
6. Mosoian, A., A. Teixeira, C.S. Burns, *et al.* 2010. Prothymosin-a inhibits HIV-1 via toll-like receptor 4-mediated type I interferon induction. *Proc. Natl. Acad. Sci. USA* **107:** 10178–10183.
7. Mosoian, A., A. Teixeira, A.A. High, *et al.* 2006. Novel function of prothymosin alpha as a potent inhibitor of human immunodeficiency virus type 1 gene expression in primary macrophages. *J. Virol.* **80:** 9200–9206.
8. Bozza, S., R. Gaziano, P. Bonifazi, *et al.* 2007. Thymosin alpha1 activates the TLR9/MyD88/IRF7-dependent murine cytomegalovirus sensing for induction of anti-viral responses in vivo. *Int. Immunol.* **19:** 1261–70.
9. Segal, B.H. & T. J. Walsh. 2006. Current approaches to diagnosis and treatment of invasive Aspergillosis. *Am. J. Respir. Crit. Care Med.* **173:** 707–717.
10. Romani, L., F. Bistoni, K. Perruccio, *et al.* 2006. Thymosin α1 activates dendritic cell tryptophan catabolism and established a regulatory environment for balance of inflammation and tolerance. *Blood* **108:** 2265–2274.
11. Grottesi, A., M. Sette, A.T. Palamara, *et al.* 1998. The conformation of peptide thymosin a1 in solution and in a membrane-like environment by circular dichroism and NMR spectroscopy. A possible model for its interaction with the lymphocyte membrane. *Peptides* **19:** 1731–1738.
12. Czisch, M., M. Schleicher, S. Hörger, W. Voelter & T.A. Holak. 1993. Conformation of thymosin beta 4 in water determined by NMR spectroscopy. *Eur. J. Biochem.* **218:** 335–344.
13. Stoll, R., W. Voelter & T.A. Holak. 1997. Conformation of thymosin beta 9 in water/fluoroalcohol solution determined by NMR spectroscopy. *Biopolymers* **41:** 623–634.
14. Elizondo-Riojas, M.-A., S.M. Chamow, C.W. Tuthill, *et al.* 2011. NMR structure of human thymosin alpha-1. *Biochem. Biophys. Res. Commun.* **416:** 356–361.
15. Hannappel, E., S. Davoust & B.L. Horecker. 1982. Thymosins β8 and β9: two new peptides isolated from calf thymus homologous to thymosin β4. *Proc. Natl. Acac. Sci. U.S.A.* **79:** 1708–1711.
16. Hutchinson, E.G. & J.M. Thornton. 1994. A revised set of potentials for beta turn formation in proteins. *Prot. Sci.* **3:** 2207–2216.
17. Zarbock, J., H. Oschkinat, E. Hannapel, *et al.* 1990. The solution conformation of thymosin b4: a nuclear magnetic resonance and simulated annealing study. *Biochemistry* **29:** 7814–7821.
18. Richardson, J.S. & D.C. Richardson. Amino acid preferences for specific amino acids at the ends of alpha helices. 1988. *Science* **240:** 1648–1652.
19. Seale, J.W., S. Rajgopal & G.D. Rose. 1994. Sequence determinants of the capping box, a stabilizing motif af the N-termini of alpha helices. *Protein Sci.* **3:** 1741–1745.
20. Aurora, R., R. Srinivasan & G.D. Rose. 1994. Rules for α-helix termination by glycine. *Science* **264:** 1126–1130.
21. Schellman, J.A. 1958. The factors affecting the stability of hydrogen-bonded polypeptide structures in solution. *J. Phys. Chem.* **62:** 1485–1494.
22. Serrano, L. & A.R. Fersht. 1989. Capping and alpha-helix stability. *Nature* **342:** 296–299.
23. Safer, D., M. Elzinga & V.T. Nachmias. 1991. Thymosin beta 4 and Fx, an actin-sequestering peptide, are indistinguishable. *J. Biol. Chem.* **266:** 4029–4032.
24. Weber, A., V.T. Nachmias, C.R. Pennise, *et al.* 1992. Interaction of thymosin beta 4 with muscle and platelet actin: implications for actin sequestration in resting platelets. *Biochemistry* **31:** 6179–6185.
25. Cassimeris, L., D. Safer, V.T. Nachmias & S.H. Zigmond. 1992. Thymosin beta 4 sequesters the majority of G-actin in resting human polymorphonuclear leukocytes. *J. Cell. Biol.* **119:** 1261–1270.
26. Paunola, E., P.K. Mattila & P. Lappalainen. 2002. WH2 domain: a small, versatile adapter for actin monomers. *FEBS Lett.* **513:** 92–97.
27. Holmes, K.C., D. Popp, W. Gebhard & W. Kabsch. 1990. Atomic model of the actin filament. *Nature* **347:** 44–49.
28. McLaughlin, P.J., J.T. Gooch, H.G. Mannherz & A.G. Weeds. 1993. Structure of gelsolin segment 1-actin complex and the mechanism of filament severing. *Nature* **364:** 685–692.
29. Schutt, C.E., J.C. Myslik, M.D. Rozycki, *et al.* 1993. The structure of crystalline profiling-β-actin. *Nature* **365:** 810–816.
30. Domanski, M., M. Hertzog, J. Coutant, *et al.* 2004. Coupling of folding and binding of thymosin β4 upon interaction with monomeric actin monitored by nuclear magnetic resonance. *J. Biol. Chem.* **279:** 23637–23645.

Ann. N.Y. Acad. Sci. ISSN 0077-8923

ANNALS OF THE NEW YORK ACADEMY OF SCIENCES
Issue: *Thymosins in Health and Disease*

Fragments of β-thymosin from the sea urchin *Paracentrotus lividus* as potential antimicrobial peptides against staphylococcal biofilms

Domenico Schillaci,[1] Maria Vitale,[2] Maria Grazia Cusimano,[1] and Vincenzo Arizza[3]

[1]Department of Molecular and Biomolecular Science and Technology (STEMBIO), University of Palermo, Palermo, Italy. [2]Department of Molecular Biology, Instituto Zooprofilattico Sperimentale Sicilia, Palermo, Italy. [3]Department of Environmental Biology and Biodiversity, University of Palermo, Palermo, Italy

Address for correspondence: Domenico Schillaci, Department of Molecular and Biomolecular Science and Technology STEMBIO, Università degli Studi di Palermo, Via Archirafi, 32-90123 Palermo, Italy. domenico.schillaci@unipa.it

The immune mediators in echinoderms can be a potential source of novel antimicrobial peptides (AMPs) applied toward controlling pathogenic staphylococcal biofilms that are intrinsically resistant to conventional antibiotics. The peptide fraction <5 kDa from the cytosol of coelomocytes of the sea urchin *Paracentrotus lividus* (5-CC) was tested against a group of Gram-positive and Gram-negative pathogen reference strains. The 5-CC of *P. lividus* was active against all planktonic-tested strains but also showed antibiofilm properties against staphylococcal strains. Additionally, we demonstrated the presence of three small peptides in the 5-CC belonging to segment 9-41 of a *P. lividus* β-thymosin. The smallest of these peptides in particular, showed the common chemical–physical characteristics of AMPs. This novel AMP from β-thymosin has high potential activity as an antibiofilm agent, acting on slow-growing bacterial cells that exhibit a reduced susceptibility to conventional antibiotics and represent a reservoir for recurrent biofilm-associated infections.

Keywords: antibiofilm agents; antimicrobial peptides (AMPs); staphylococcal biofilms

Introduction

Staphylococci can induce a wide spectrum of infectious diseases associated with remarkable morbidity and mortality.[1] Pathogenic staphylococci have an extraordinary ability to acquire several antibiotic resistance traits, and the rise of community and hospital-acquired methicillin-resistant *Staphylococcus aureus* (MRSA) is a major health problem worldwide. This scenario has created an urgent need for novel therapeutic approaches[2] to control drug-resistant bacterial strains, not only in their free-living planktonic form, but also when encountered as biofilms—bacterial communities able to grow on surfaces and surrounded by an extracellular polymeric substance (EPS) matrix.

The ability to form biofilms, is probably the most important virulence factor of staphylococci in the development of the chronic and persistent form of several infectious diseases in humans such as otitis media, osteomyelitis, endophtalmitis, urinary tract infections, acute septic arthritis, native valve endocarditis, burn or wound infections, and cystic fibrosis-associated pulmonary infections.[3–9]

Furthermore, staphylococcal biofilms are commonly isolated from medical device-related infections with *S. aureus* mainly involved in metal-biomaterial infections, while *Staphylococcus epidermidis* is more often observed in polymer-associated infections.[10] The Gram-positive pathogens *S. aureus*, *S. epidermidis*, and *Enterococcus faecalis* represent more than 50% of the species isolated from patients with medical device–associated infections,[11] and catheter-related bloodstream infections (CRBSIs) during intensive care unit (ICU) stays in four European countries (France, Germany, Italy,

doi: 10.1111/j.1749-6632.2012.06652.x

UK) have an estimated cost of € 163.9 million in health care.[12]

Staphylococcal biofilm resistance to antibiotics: a multifactorial mechanism

Bacterial biofilms are more resistant to host immune defense systems and display a significantly high degree of antibiotic tolerance.[13] Antibiotic resistance in biofilms is multifactorial, because biofilm structured bacteria develop different mechanisms of resistance. The bacteria in the external layers of the community, for example, are more active in cell division and energetic metabolism compared to the internal layers, due to oxygen and nutrient gradients from the top to the bottom of a biofilm. A metabolically heterogeneous bacterial population differs markedly from a free-living (planktonic) population[14] and nutrient-depleted zones can result in stationary phase-like cells (dormant metabolic state) with reduced susceptibility to antibiotics.[15]

The EPS matrix may retard the rate of penetration of antibiotics enough to induce the expression of genes that mediate resistance within the biofilm.[3] Other well-known mechanisms such as the production of enzymes that degrade antibiotics, alteration of targets, or overexpression of efflux pumps that have a broad range of substrates, are associated with the planktonic cells, but bacterial cells growing in biofilm increase horizontal gene transmission so they can easily spread antibiotic resistance traits.[16] Furthermore, specialized populations of persister cells in the *S. aureus* biofilms, remain in a dormant state in the presence of an antibiotic, with no growth and no death.[17] This mechanism is believed to be responsible for recurrent infections in hospital settings because the persister cells give rise to a normal bacterial colony after drug removal.

It has been observed that *S. aureus* in biofilms is 100–1000 times less susceptible to antibiotics than equivalent populations of planktonic bacteria.[18] Conventional antibiotics can be effective against metabolically active bacterial cells but currently no effective therapies for staphylococcal biofilms exist. Early removal of the device or surgical intervention, remain the most effective means to treat biofilm-associated infections, to date.[4] Therefore, there is an urgent need for novel treatments, strategies, and antistaphylococcal biofilm agents.

Discovery of novel antistaphylococcal biofilm agents

In the biofilm preclinical research field, three different approaches are primarily followed: screening of novel compounds (synthetic or natural) that inhibit staphylococcal biofilms through direct effects on bacterial growth and viability; target-based strategy for discovering agents that show antibiofilm properties by targeting specific pathways essential for staphylococcal biofilm formation; and enzymes that target staphylococcal biofilm matrix.[19,20] This paper will focus on discovery of novel antimicrobial peptides (AMPs) derived from the beta-thymosin peptide of *Paracentrotus lividus* as new antistaphylococcal biofilm agents.

Activity-based screening of antibiofilm agents

We focused on the immune system of marine invertebrates as a relatively underexplored source of new antimicrobial agents. Echinoderms are intertidal benthic organisms that are constantly exposed to a persistent threat of infection by high concentrations of bacteria and viruses from the marine environment. The survival of these organisms depends on efficient antimicrobial mechanisms that protect them against pathogens. We focused on the coelomocytes, the immune mediators in echinoderms. In particular, our study focused on the sea urchin, *Paracentrotus lividus,* which is a common species in the Mediterranean sea. The coelomocytes of echinoderms are responsible of a wide repertoire of cellular and humoral immunologic functions, including cellular recognition, phagocytosis, cytotoxicity, and the production of antimicrobial peptides (AMPs).[21,22] In addition to AMPs, a 60-kDa protein, which showed antibacterial activity, has been isolated from lysates of coelomocytes from *P. lividus*.[23] The survival and fitness of *P. lividus* in marine environments suggest that its innate immune system is potent and effective, since this species is a long-living organism. Moreover, it lives in an infralitoral environment where it is exposed to pathogenic attacks from invading microorganisms also of anthropic origin, and it is not fouled, so it clearly has developed strategies to prevent bacterial colonization on its surface. All these biological and ecological aspects render the sea urchin *P. lividus* a good source for AMPs with high potential as novel antimicrobial molecules.

Antimicrobial peptides (AMPs) are characterized by a small molecular size (<10 kDa, or ~10–50 amino acids) and broad antibacterial activity. A large variety of AMPs have been described in marine invertebrates and represent the major humoral defense system against pathogens: defensin, mytícin, and mytilin in mussels; penaeidin in shrimp; tachyplesin; and polyphemusin in horseshoe crab; clavanin; and styelin in ascidians, and Ci-PAP-A22 in *Ciona intestinalis*.[24–32]

AMPs in *P. lividus* have not been previously evaluated. We studied the antimicrobial and antistaphylococcal biofilm activity of a 5-kDa peptide fraction from coelomocytes cytosol (5-CC) against reference strains and isolates of human and animal origin.

Biological activity of a 5-kDa peptide fraction from coelomocytes cytosol

The 5-CC of *P. lividus* was tested against a Gram-positive (*S. aureus* and *S. epidermidis*, including drug-resistant strains) group and a Gram-negative (*Pseudomonas aeruginosa*, *E. coli*) group, and planktonic reference yeast (*C. albicans*, *C. tropicalis*) strains by using a microdilution method and determining the minimum inhibitory concentration (MIC). The 5-CC showed a broad antimicrobial activity against all tested strains (Table 1). Moreover, 5-CC showed antibiofilm properties against staphylococcal biofilms of reference strains *S. epidermidis* DSM 3269 and *S. aureus* ATCC 29213. The antimicrobial efficacy of 5-CC against biofilms of the clinical strain *S. epidermidis* 1457 was also tested using live/dead staining in combination with confocal laser scanning microscopy. At a sub-MIC concentration (31.7 mg/mL) of 5-CC the formation of young (six hours old) and mature (24-hours old) staphylococcal biofilms was inhibited. We observed an interesting inhibitory effect of 5-CC at a sub-MIC concentration, either on the formation of a young biofilm (six hours old) of *S. epidermidis* 1457 or on the formation of a mature biofilm (24 hours old) of the same clinical strain (Fig. 1).[36] The susceptibility to antimicrobial treatment of a biofilm can depend on the stage of development (age) of the biofilm itself. A mature biofilm can be more tolerant to antimicrobial treatment than a young biofilm. Live/dead staining was used to assay bacterial viability with or without treatment. As there was no sign of dead cells, the reduction in bacterial adhesion could be due to an interference of 5-CC peptides with microbial surface proteins (adhesins, autolysins) that facilitate attachment to surfaces in the first step of staphylococcal biofilm formation.[33,34]

Table 1. **Antimicrobial activity of a 5 kDa peptide fraction from celomocytes (5-CC)[a]**

	5-CC MIC values in mg/mL
S. aureus ATCC 29213	126.8
S. aureus ATCC 25923	63.4
S. aureus ATCC 43866	63.4
S. epidermidis 1457	126.8
S. epidermidis DSM 3269	253.7
E. coli ATCC 25922	126.8
P. aeruginosa ATCC 9027	253.7
C. albicans ATCC 10231	31.7
C. tropicalis ATCC 13813	15.8

[a]Values *in vitro* expressed in mg/mL for all strains tested.

Considering that biofilms can be found in virtually all natural ecosystems that support microbial growth, they also have enormous impact in veterinary medicine because they can be responsible for the failure in antimicrobial therapy in bacterial infections or in a failure of properly sanitizing of food processing plants. Many aspects of the farm management, in animal health and welfare and in food processing premises should be reconsidered in the light of these very common bacterial communities. *S. aureus* is a major pathogen of mastitis, which is one of the most common diseases in dairy cattle, and we are currently testing the 5-CC against some *S. aureus* isolates of animal origin. From preliminary experiments, we determined good antistaphylococcal activity against planktonic *S. aureus* isolates (MIC values ranging from 0.62 to 0.31 μg/mL). We

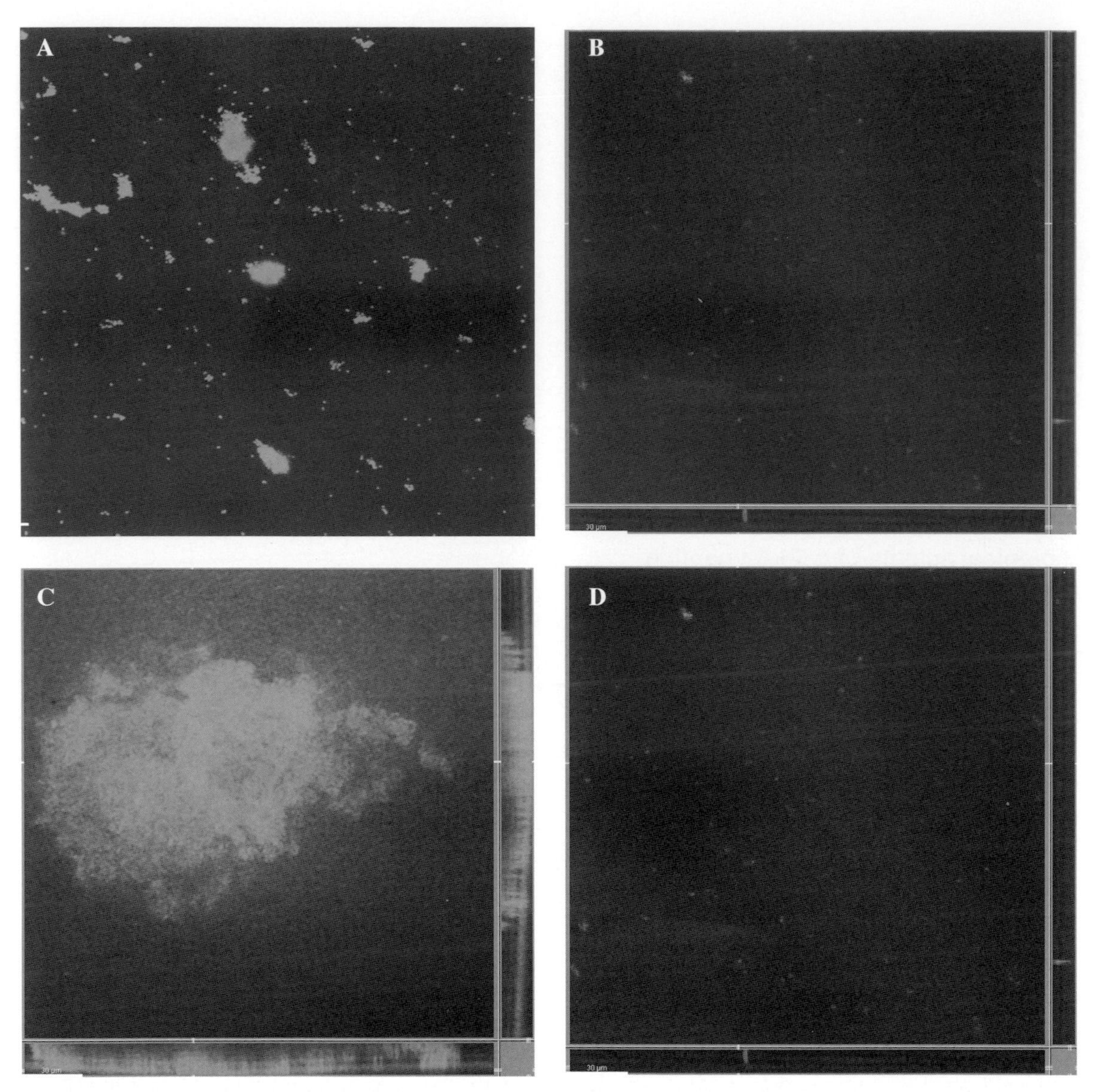

Figure 1. Preventative inhibitory activity of 5-CC. (Panel A) *Staphylococcus epidermidis* 1457 growth control (six hours old); (panel B) treated with a concentration of 31.7 mg/mL after six hours; (panel C) *S. epidermidis* 1457 growth control (24 hours old); (panel D) treated with a concentration of 31.7 mg/mL after 24 hours. After 6- or 24-h treatment, the biofilms were stained with live/dead materials (SYTO9, green; PI, red) and observed using CLMS. The assays were repeated at least twice, and similar results were obtained.[36]

also plan to evaluate the activity of 5-CC against preformed 24-h-old biofilms of staphylococcal isolates of veterinary importance.

AMPs from β-thymosin of *P. lividus*

The antimicrobial defense system of marine invertebrates is based solely on an innate immune system that includes both humoral and cellular responses. AMPs constitute a major component of their humoral immunity. They are short cationic, amphipathic sequences of amino acids ranging around 10–50 amino acids in length. Marine invertebrate AMPs display broad antimicrobial spectra, even against human pathogens.[35]

In our experimental work, we observed that 5-CC possessed a broad antimicrobial activity against all tested pathogens. Small-sized molecules with a broad antimicrobial spectrum are two common characteristics of AMPs, hence we employed RP-HPLC / nESI-MSMS to confirm the presence of AMPs in the 5-CC content. Three principal peptides in 5-CC, whose molecular weights were respectively 1251.7, 2088.1, and 2292.2 Da, were identified: the (9–19), (12–31), and (24–41) fragments of a β-thymosin of *P. lividus* (NCBInr acc.no/gi/22474470) whose molecular mass is 4592 Da.[36] We found by BLAST analysis that β-thymosin of *P. lividus* has an identity of 87% with human β-thymosin 10.

The β-thymosins are a family of highly conserved polar 5-kDa peptides originally thought to be thymic hormones. They are present at high concentrations in almost every cell from vertebrate phyla and have several biological functions due to direct and indirect effects on the actin cytoskeleton. There is little information about the function of thymosins in invertebrates, but their presence has been reported in marine invertebrates[37,38] and in insects where they are upregulated by microbial infections.[39]

By analyzing some important chemical–physical properties, such as hydrophobicity, charge, and presence of hydrophobic residues on the same not polar face, we found that the smallest fragment, fragment 1, (9–19 of β-thymosin), 11 amino acids in length, has a good chance of being an antimicrobial peptide:[40] it has a net positive charge because of an excess number of lysine residues, and it has three hydrophobic residues on the same face and a total hydrophobic ratio of 36%. Hydrophobic and charged residues may permit interaction with bacterial membranes.[41] Moreover fragment 1 of β-thymosin has an alpha-helix structure, the most common structure of AMPs in nature, and has a similarity with already described AMPs produced by a variety of organisms, for instance, a similarity of 35% with the Jelleine III found in royal jelly of honeybees (*Apis mellifera*).[42]

Fragment 2 (12–31 of β-thymosin), 20 amino acids in length, has a positive charge and a similarity of 40.9% with maculatin, a peptide obtained from skin glands of the tree frog *Litoria genimaculate.*[43] Fragment 3 (24–41 of β-thymosin), and β-thymosin itself, are negatively charged and have little chance to be AMPs.

Interestingly, the entire sequence from amino acid 9 to amino acid 41, may form alpha helices and have at least five residues on the same hydrophobic face. This region may interact with membranes and has also good chance to be an AMP; moreover it has a similarity of 39% with latarcins, antimicrobial and cytolytic peptides from the venom of the spider *Lachesana tarabaevi* (Table 2).[44]

AMPs from β-thymosin of *P. lividus* as potential antibiofilm agents

The tolerance of biofilms to antibiotics is mainly due to the slow growth and low metabolic activity

Table 2. Chemical–physical properties of *P. lividus* β-thymosin fragments and similarity with already described AMPs

Peptide	MW	Total net charge	Percentage hydrophobic residues	Sequence and hydrophobic residues (underlined) on the same face	Similarity (> 35%)
Fragment 1 (9–19)	1251.7	+1	36%	E V A S F D K S K L K	Jelleine III
Fragment 2 (12–31)	2293.2	+2	20%	S F D K S K L K K A E T Q E K N T L P T	Maculatin
Fragment 3 (24–41)	2088.1	−1	16%	Q E KN T L P T K ET I EQE K T A	
Entire sequence (9–41)	3745.1	0	24%	E V A S F D K S K L K K A E T Q E K N T L P T K E T I E Q E K T A	Latarcin
β-thymosin (1–41)	4583.1	−1	26%	MADKPDVSEVASFDKSKLKKAE TQEKNTLPTKETIEQEKTA	

of bacteria in such communities, so they are intrinsically resistant to antibiotics, such as β-lactams, which target dividing and metabolically active cells. On the contrary, the prevalent mechanism of action of AMPs is due to their ability to permeabilize and/or to form pores within the cytoplasmic membranes, so they have a high potential to act also on slow-growing or even nongrowing bacteria that exhibit a reduced susceptibility to many antibiotics and represent a reservoir for recurrent biofilm infections. The AMPs also have a high potential for inhibiting biofilm formation, in fact, they can act at several stages of biofilm formation and with different mechanisms of action: they may minimize initial adhesion of microbial cells to abiotic surfaces by altering the adhesive features of plastic surfaces, or by binding to microbial surfaces via electrostatic interactions, or may prevent biofilm maturation by killing the early surface colonizers, or by inhibiting quorum sensing (QS), that is, the communication system used by many bacteria to build a biofilm.[45]

Staphylococcal biofilms are responsible for many biomaterial associated infections (BAI), including persistent forms of some infectious diseases in humans. The continual increase in the use of medical devices is associated with a significant risk of infectious complications, including blood stream infections, septic thrombophlebitis, endocarditis, metastatic infections, and sepsis.[46,47] Biofilm associated infections of indwelling medical devices are usually resolved after replacement of the device but involve a prolonged hospital stay and increased healthcare costs. Considering also that increasing numbers of elderly patients require indwelling medical devices like artificial knees and hips, a new generation of antiinfective agents effective in the prevention or eradication of biofilms is needed.[48]

AMPs derived from β-thymosin of *P. lividus* for their chemical–physical characteristics and predicted activity are attractive candidates for potential therapeutic development in medical and veterinary field. Our current experimental work is aimed to confirm the predicted activity of the fragments of β-thymosin and to improve their potential as novel effective chemical countermeasures against staphylococcal biofims.

Conflicts of interest

The authors declare no conflicts of interest.

References

1. Tang, Y.W. & C.W. Stratton. 2010. *Staphylococcus aureus*: an old pathogen with new weapons. *Clin. Lab. Med.* **30:** 179–208.
2. Ohlsen, K. & U. Lorenz. 2007. Novel targets for antibiotics in *Staphylococcus aureus*. *Future Microbiol.* **2:** 655–666.
3. Hall-Stoodley, L. & P. Stoodley. 2009. Evolving concepts in biofilm infections. *Cellular Microbiology* **11:** 1034–1043.
4. Brady, R.A., J.G. Leid, J.H. Calhoun, J.W. Costerton & M.E. Shirtliff. 2008. Osteomyelitis and the role of biofilm in chronic infection. *FEMS Immunol. Med. Microbiol.* **52:** 13–22.
5. Callegan, M.C., M.S. Gilmore, M. Gregory, *et al.* 2007. Bacterial endophthalmitis: therapeutic challenges and host-pathogen interactions. *Prog. Retin. Eye Res.* **26:** 189–203.
6. Ronald, A. 2002. The etiology of urinary tract infection: traditional and emerging pathogens. *Am. J. Med.* **113:** 14s–19s.
7. Shirtliff, M.E. & J.T. Mader. 2002. Acute septic arthritis. *Clin. Microbiol. Rev.* **15:** 527–544.
8. Donlan, R.M. & J.W. Costerton. 2002. Biofilms: survival mechanisms of clinically relevant microorganisms. *Clin. Microbiol. Rev.* **15:** 167–193.
9. Davies, J.C. & D. Bilton. 2009. Bugs, biofilms, and resistance in cystic fibrosis. *Respir. Care* **54:** 628–640.
10. Götz, F. 2002. *Staphylococcus* and biofilms. *Molecular Microbiology* **43:** 1367–1378.
11. Donelli, G., I. Francolini, D. Romoli, *et al.* 2007. Synergistic activity of dispersin B and cefamandole nafate in inhibition of staphylococcal biofilm growth on polyurethanes. *Antimicrob. Agents Chemother.* **51:** 2733–2740.
12. Tacconelli, E., G. Smith, K. Hieke, A. Lafuma & P. Bastide. 2009. Epidemiology, medical outcomes and costs of catheter-related bloodstream infection in intensive care units of four European countries: literature and registry-based estimates. *J. Hosp. Infect.* **72:** 97–103.
13. Høiby, N., T. Bjarnsholt, M. Givskov, S. Molin & O. Ciofu. 2010. Antibiotic resistance of bacterial biofilms. *Int.J. Antimicrob. Agents* **35:** 322–332.
14. Costerton, J.W., Z. Lewandowski, D.E. Caldwell, D.R. Korber & H.M. Lappin-Scott. 1995. Microbial biofilms. *Annu. Rev. Microbiol.* **49:** 711–745.
15. Keren, I., N. Kaldalu, A. Spoering, Y. Wang & K. Lewis. 2004. Persister cells and tolerance to antimicrobials. *FEMS Microbiol. Lett.* **230:** 13–18.
16. Molin, S. & T. Tolker-Nielsen. 2003. Gene transfer occurs with enhanced efficiency in biofilms and induces enhanced stabilisation of the biofilm structure. *Curr. Opin. Biotechnol.* **14:** 255–261.
17. Lewis, K. 2007. Persister cells, dormancy and infectious disease. *Nat. Rev. Microbiol.* **5:** 48–56.
18. Gilbert, P., D.G. Allison & A.J. McBain. 2002. Biofilms in vitro and in vivo: do singular mechanisms imply cross resistance? *J. Appl. Microbiol.* **92:** 98S–110S.
19. Schillaci, D., 2011. Staphylococcal biofilms: challenges in the discovery of novel anti-infective agents. *Microbial. Biochem. Technol.* **3:** iv–vi.

20. Kiedrowski, M.R. & A.R. Horswill. 2011. New approaches for treating staphylococcal biofilm infections. *Ann. N.Y. Acad. Sci.* **1241:** 104–121.
21. Arizza, V., F.T. Giaramita, D. Parrinello, M. Cammarata and N. Parrinello. 2007. Cell cooperation in coelomocyte cytotoxic activity of *Paracentrotus lividus* coelomocytes. *Comp. Biochem. Physiol. A Mol. Integr. Physiol.* **147:** 389–394.
22. Huang, T., A.K. Kjuul, O.B. Styrvold, *et al.* 2002. Antibacterial activity in *Strongylocentrotus droebachiensis* (Echinoidea), *Cucumaria frondosa* (Holothuroidea) and *Asterias rubens*(Asteroidea). *J. Invertebrat. Pathol.* **81:** 85–94.
23. Stabili, L., P. Pagliara & P. Roch. 1996. Antibacterial activity in the coelomocytes of the sea urchin Paracentrotuslividus. *Comp. Biochem. Physiol. B Biochem. Mol. Biol.* **113:** 639–644.
24. Boman, H. 1995. Peptide antibiotics and their role in innate immunity. *Annu. Rev. Immunol.* **13:** 61–92.
25. Cellura, C., M. Toubiana, N. Parrinello & P. Roch. 2007. Specific expression of antimicrobial peptide and HSP70 genes in response to heat-shock and several bacterial challenges in mussels. *Fish Shellfish Immunol.* **22:** 340–350.
26. Hubert, F., T. Noël & Ph. Roch. 1996. A new member of the arthropod defensin family from edible Mediterranean mussels (*Mytilus galloprovincialis*). *Eur. J. Biochem.* **240:** 302–306.
27. Mitta, G.F., T. Hubert, B. Noël & P. Roch. 1999. Myticin, a novel cysteine-rich antimicrobial peptide isolated from hemocytes and plasma of the mussel Mytilus galloprovincialis. *Eur. J. Biochem.* **265:** 71–78.
28. Mitta, G.F., F. Vandenbulcke, M. Hubert, M. Salzet & P. Roch. 2000. Involvement of mytilins in mussel antimicrobial defense. *J. Biol. Chem.* **275:** 12954–12962.
29. Destoumieux, D., M. Munoz, P. Bulet & E. Bachère. 2000. Penaeidins, a family of antimicrobial peptides from penaeid shrimp (Crustacea, Decapoda). *Cell Mol. Life Sci.* **57:** 1260–1271.
30. Miyata, T., F. Tokunaga, T. Yoneya, *et al.* 1989. Antimicrobial peptides, isolated from horseshoe crab hemocytes, tachyplesin II, and polyphemusins I and II: chemical structures and biological activity. *J. Biochem.* **106:** 663–668.
31. Lee, I.H., Y. Cho & R.I. Lehrer. 1997. Styelins, broadspectrum antimicrobial peptides from the solitary tunicate, Styela clava. *Comp. Biochem. Physiol. B Biochem. Mol. Biol.* **118:** 515–521.
32. Fedders, H. & M.A. Leippe. 2008. Reverse search for antimicrobial peptides in Ciona intestinalis: identification of a gene family expressed in hemocytes and evaluation of activity. *Dev. Comp. Immunol.* **32:** 286–298.
33. Patti, J.M. & M. Hook. 1994. Microbial adhesins recognizing extracellular matrix macromolecules. *Curr. Opin. Cell Biol.* **6:** 752–758.
34. Heilmann, C., M. Hussain, G. Peters & F. Götz. 1997. Evidence for autolysin-mediated primary attachment of *Staphylococcus epidermidis* to a polystirene surface. *Mol. Microbiol.* **20:** 1013–1024.
35. Tincu, J.A. & S.W. Taylor. 2004. Antimicrobial peptides from marine invertebrates. *Antimicrob. Agents Chemother.* **48:** 3645–3654.
36. Schillaci, D., V. Arizza, N. Parrinello, *et al.* 2010. Antimicrobial and antistaphylococcal biofilm activity from the sea urchin *Paracentrotus lividus*. *J. Appl. Microbiol.* **108:** 17–24.
37. Safer, D., R. Golla & V.T. Nachmias. 1990. Isolation of a 5-kilodalton actin sequestering peptide from human blood platelets. *Proc. Natl. Acad. Sci. USA* **87:** 2536–2540.
38. Romanova, E.V., M.J. Roth, S.S. Rubakhin, *et al.* 2006. Identification and characterization of homologues of vertebrate β-thymosin in the marine mollusk *Aplysia californica*. *J. Mass. Spectrom.* **41:** 1030–1040.
39. Zhang, F.X., H.L. Shao, J.X. Wang & X.F. Zhao. 2011. β-Thymosin is upregulated by the steroid hormone 20-hydroxyecdysone and microorganisms. *Insect. Mol. Biol.* **20:** 519–527.
40. Wang, Z. & G. Wang. 2004. ADP: the antimicrobial peptide database. *Nucleic Acids Res.* **32:** D590–D592.
41. Hancock, R.E.W., K.L. Brown & N. Mookherjee. 2006. Host defense peptides from invertebrates-emerging antimicrobial strategies. *Immunobiology* **211:** 315–322.
42. Fontana, R., M.A. Mendes, B.M. de Souza, *et al.* 2004. Jelleines: a family of antimicrobial peptides from the Royal Jelly of honeybees (*Apis mellifera*). *Peptides* **25:** 919–928.
43. Rozek, T., R.J. Waugh, S.T. Steinborner, *et al.* 1998. The maculatin peptide from the skin glands of the tree frog *Litoria genimaculata*. A comparison of the structures and antibacterial activities of maculatin 1.1 and caerin 1.1. *J. Pept. Sci.* **4:** 111–115.
44. Kozlov, S.A., A.A. Vassilevski, A.V. Feofanov, *et al.* 2006. Latarcins, antimicrobial and cytolytic peptides from the venom of the spider *Lachesana tarabaevi* (Zodariidae) that exemplify biomolecular diversity. *J. Biol. Chem.* **281:** 20983–20992.
45. Batoni, G., G. Maisetta, F.L. Brancatisano, S. Esin & M. Campa. 2011. Use of antimicrobial peptides against microbial biofilms: advantages and limits. *Curr. Med. Chem.* **18:** 256–279.
46. Donelli, G., P. De Paoli, G. Fadda, *et al.* 2001. A multicenter study on central venous catheter-associated infections in Italy. *J. Chemother.* **13:** 251–262.
47. Parsek, M.R. & P.K. Singh. 2003. Bacterial biofilms: an emerging link to disease pathogenesis. *Annu. Rev. Microbiol.* **57:** 677–701.
48. Lynch, A.S. & D. Abbanat. 2010. New antibiotic agents and approaches to treat biofilm-associated infections. *Expert Opin. Ther. Pat.* **20:** 1373–1387.

Ann. N.Y. Acad. Sci. ISSN 0077-8923

ANNALS OF THE NEW YORK ACADEMY OF SCIENCES
Issue: *Thymosins in Health and Disease*

Development of an analytical HPLC methodology to study the effects of thymosin β4 on actin in sputum of cystic fibrosis patients

Mahnaz Badamchian,[1] Ali A. Damavandy,[2] and Allan L. Goldstein[1]

[1]Department of Biochemistry and Molecular Biology, The George Washington University School of Medicine and Health Sciences, Washington DC. [2]Department of Internal Medicine, Beth Israel Deaconess Medical Center, Harvard Medical School, Boston, Massachusetts

Address for correspondence: Mahnaz Badamchian, Ph.D., 10405 Chelsea Manors Court, Great Falls, VA 22066. mbadam@chemlifeinc.com

A high-performance liquid chromatography (HPLC) methodology is presented. Using bovine and rabbit F- and G-actin, this methodology results in both fractions as being well-resolved peaks, which were confirmed by dot blot immunoassay and fluorescence microscopy. F- and G-actin were incubated with thymosin β4 (Tβ4) and DNase and then analyzed by HPLC, which indicated that Tβ4 and DNase inhibit G-actin polymerization and that Tβ4 depolymerizes F-actin in a dose- and time-dependent manner. The F- and G-actin content in sputum from healthy controls and cystic fibrosis (CF) patients were measured by HPLC before and after incubation with Tβ4, DNase, and gelsolin. These data demonstrate higher quantities of F-actin in the sputum of CF patients compared to healthy individuals, and also demonstrate a significantly increased F/G-actin ratio in CF sputum. Further, Tβ4, DNase, and gelsolin each increase the depolymerization of F-actin in CF sputum in a dose-dependent fashion that is additive when these agents are combined.

Keywords: thymosin β4; actin; cystic fibrosis; sputum; HPLC

Introduction

Actin is the most abundant protein in mammalian cells, making up to 10–20% of total cellular protein. Actin is involved in many nonmuscle cellular functions, such as motility, chemotaxis, cytokinesis, and phagocytosis. Cellular regulation of these processes requires the dynamic assembly and disassembly of actin between G-actin (monomers) to F-actin (filaments and higher order cytoskeletal structures).[1–3] The reversible polymerization control of actin is highly regulated in cells by several actin-binding proteins. These proteins can regulate actin polymerization by preventing G-actin availability (sequestering), by limiting the length of actin polymers (capping), or by breaking noncovalent actin–actin bonds within a formed actin filament (severing).[2] However, disregulation of actin assembly and disassembly cycles, which occurs when actin is released from damaged cells into the extracellular space, contributes to the pathophysiology of many disease states, including septic shock, acute respiratory distress syndrome, and cystic fibrosis (CF).[2–4] Control of actin in the extracellular space, therefore, could be important in controlling the pathogenicity of these disease states.

Cystic fibrosis is the most common inherited lethal disease in people of European descent.[5] The defective transmembrane conductance regulator gene in CF patients results in abnormally thick and viscous luminal secretions throughout the body but most notably in the airways of the respiratory system, causing the majority of the morbidity and mortality of the disease. The increased viscosity and poor mobility of sputum provides a substrate for chronic inflammation, repeated infections, scarring, and, ultimately, bronchiectasis of the airways, which greatly compromises the function of the lungs. Previously, extracellular DNA and filamentous F-actin have been identified in the sputum of CF patients as

doi: 10.1111/j.1749-6632.2012.06671.x

causative agents in its abnormal biophysical properties.[6,7] The source of these components is pus, which is derived from masses of degenerating neutrophils that migrate to the site of infection and inflammation to combat bacterial infections and release their intracellular contents into affected airways. Actin released from damaged inflammatory cells and respiratory epithelia spontaneously polymerizes in airway secretions into long protease-resistant and highly visco-elastic filaments that greatly increase the stickiness of CF sputum.[7]

Thymosin β4 (Tβ4), originally isolated from a bovine thymus extract termed thymosin fraction 5, is a ubiquitous 43 amino acid peptide and the major actin-sequestering molecule in nonmuscle eukaryotic cells.[8] A major property of Tβ4 is its capability of maintaining a large pool of unpolymerized actin intracellularly.[9] Under physiological conditions, it has been shown to form a 1:1 complex with G-actin and is naturally present in large enough quantities to bind greater than 90% of G-actin in resting platelets.[10] In addition to its previously reported immune-modulatory and regenerative properties, Tβ4 can downregulate the production of several inflammatory cytokines including IL-1, IL-6, and TNF-α following the administration of LPS, and reduce the toxicity of this endotoxin.[11]

It has been reported previously that actin filament length correlates with sputum cohesivity, that Tβ4 exhibits a time- and dose-dependent decrease in sputum cohesivity in CF sputum, and that Tβ4 in combination with dornase alfa increases cough and mucociliary transportability of CF sputum *in vitro*.[12,13]

In spite of the vast research of the effect of different mucolytic agents on the actin filaments in CF sputum, there has been a paucity of direct, quantitative, and reliable methods for the measuring the F-actin content in human sputum samples. The goals of this study were to develop a HPLC methodology to directly and quantitatively measure F-actin and G-actin in CF and control sputum and to apply this technique to study the mucolytic properties of Tβ4 *in vitro*.

HPLC separation of purified G-actin, F-actin, and Tβ4

In order to determine the effect of Tβ4 on the integrity of F-actin *in vitro*, we are developing a technique using high-performance liquid chromatography (HPLC), which allows for the quantitative measurement of G-actin and F-actin in test samples and human sputum. The separation is performed on a Waters HPLC system equipped with a model 490E multiwave length detector set at 214 nm (detects Tβ4) and 280 nm (does not detect Tβ4), and a 7.5 mm × 30 cm Protein Pak 300 SW column. The running buffer was a 5 mM Tris buffer (pH 7.2) containing 0.1 M KCl, 2 mM $MgCl_2$, and 0.01% NaN_3. The column is equilibrated for 15 min before sample application, and a 15-min run time at a flow rate of 1 mL/min was used with collection of 1-min fractions and a time to salt peak of 11 minutes. We additionally perform a dot blot immunoassay to examine the immunoreactivity of the HPLC fractions and CF sputum to confirm their actin content. Our application of the dot blot data indicates that monoclonal antiactin antibody reliably stains purified G-actin and F-actin as well as CF sputum, yet is specific and does not stain thymic extracts such as thymoject, thymosin fraction 5, or the purified thymic peptides thymosin α1 (Tα1) and Tβ4 (data not shown).

HPLC separation of purified rabbit actin in the presence of Tβ4 indicated resolved peaks for F-actin and G actin, which were dot blot positive, as well as Tβ4 and salt, which were not immune reactive (Fig. 1, preliminary data). Therefore, using our newly developed HPLC methodology, we reliably separate Tβ4, F-actin, and G-actin peaks and confirm the content of these peaks using the dot blot immunoassay.

Effect of Tβ4 on polymerization and depolymerization of actin

In addition to quantifying the actin content of test samples, HPLC provides the capability of performing time- and concentration-dependent kinetic studies of actin polymerization and depolymerization in the presence of Tβ4. Incubating a fixed concentration of G-actin (20 μM) with different concentrations of Tβ4 (40 μM and 100 μM), followed by HPLC analysis, demonstrated a dose-dependent relationship in the ability of Tβ4 to inhibit the polymerization of G-actin monomers into F-actin filaments (Fig. 2A). Similarly, incubating F-actin (20 μM) both in the presence and absence of Tβ4 demonstrates the depolymerization of F-actin into the globular G-actin monomers by Tβ4 (Fig. 2B). These data indicate that Tβ4 regulates actin

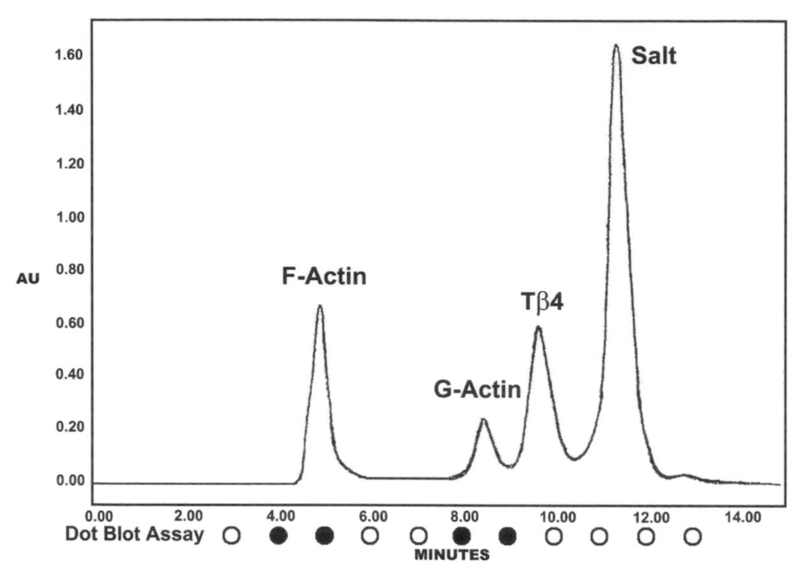

Figure 1. HPLC separation of Tβ4, F-actin, and G-actin. Tβ4 (100 μg) and rabbit actin (100 μg) were applied to a 7.5 mm x 30 cm protein pak 300 SW column. One milliliter fractions were collected and assayed for immunoreactivity using the antiactin antibody dot blot assay. Detection was at 214 nm. Data represent work in progress.

activity by two distinct mechanisms: first by inhibiting the polymerization of G-actin into F-actin and, secondly, by promoting the depolymerization of F-actin into G-actin. Finally, incubation of F-actin with Tβ4 at a molar ratio of 1:5 (actin/Tβ4) with HPLC analysis performed at several time points over 120 min demonstrates the time-dependent action of Tβ4 on F-actin with near-complete depolymerization of F-actin achieved at 90 min (Fig. 2C). Of note, Tβ4 peaks are not seen in Figures 2A–C as detection is at 280 nm.

Using this newly developed HPLC methodology, we have also studied the effect of DNase on G-actin polymerization. Incubation of G-actin with DNase followed by HPLC analysis demonstrates inhibition of G-actin polymerization (data not shown).

Effect of Tβ4 on the content of F-actin in CF sputum

The principles of the above studies performed on the purified actin protein were next applied to human sputum samples. Assessment of the actin content of control and CF sputum was performed *in vitro* using HPLC. Expectorated sputum collected from CF patients, as well as healthy controls were processed and applied to the HPLC column and fractions were collected and tested by the above-described dot blot immunoassay with resolution of distinct F-actin and G-actin peaks (Fig. 3A). Calculation of peak area measurements shows an increase in the F-actin content of CF sputum compared to controls, as well as a substantial increase in the fraction of total actin polymerized in the filamentous F-actin form in CF sputum (Fig. 3B).

The effect of Tβ4 on the content of F-actin in CF sputum was also studied using HPLC. CF patient sputum samples were processed, and 150 μL aliquots were incubated either in vehicle alone or in the presence of 500 μg of Tβ4 for 60 min and analyzed by HPLC. CF sputum incubated in the presence of Tβ4 exhibited a statistically significant and substantial 36% decrease in the content of F-actin ($P = 0.003$) (Fig. 3C).

Effect of different mucolytic agents on F-actin depolymerization in CF sputum

The effect of Tβ4 on F-actin depolymerization in CF sputum was compared to other known mucolytic agents such as gelsolin, a potent actin-severing protein, and DNase, a selective DNA cleaving enzyme whose recombinant human form is used clinically in the treatment of cystic fibrosis patients. CF sputum samples were incubated with differing molar ratios (expressed as moles of test agent per moles of F-actin) of DNase, gelsolin, and Tβ4

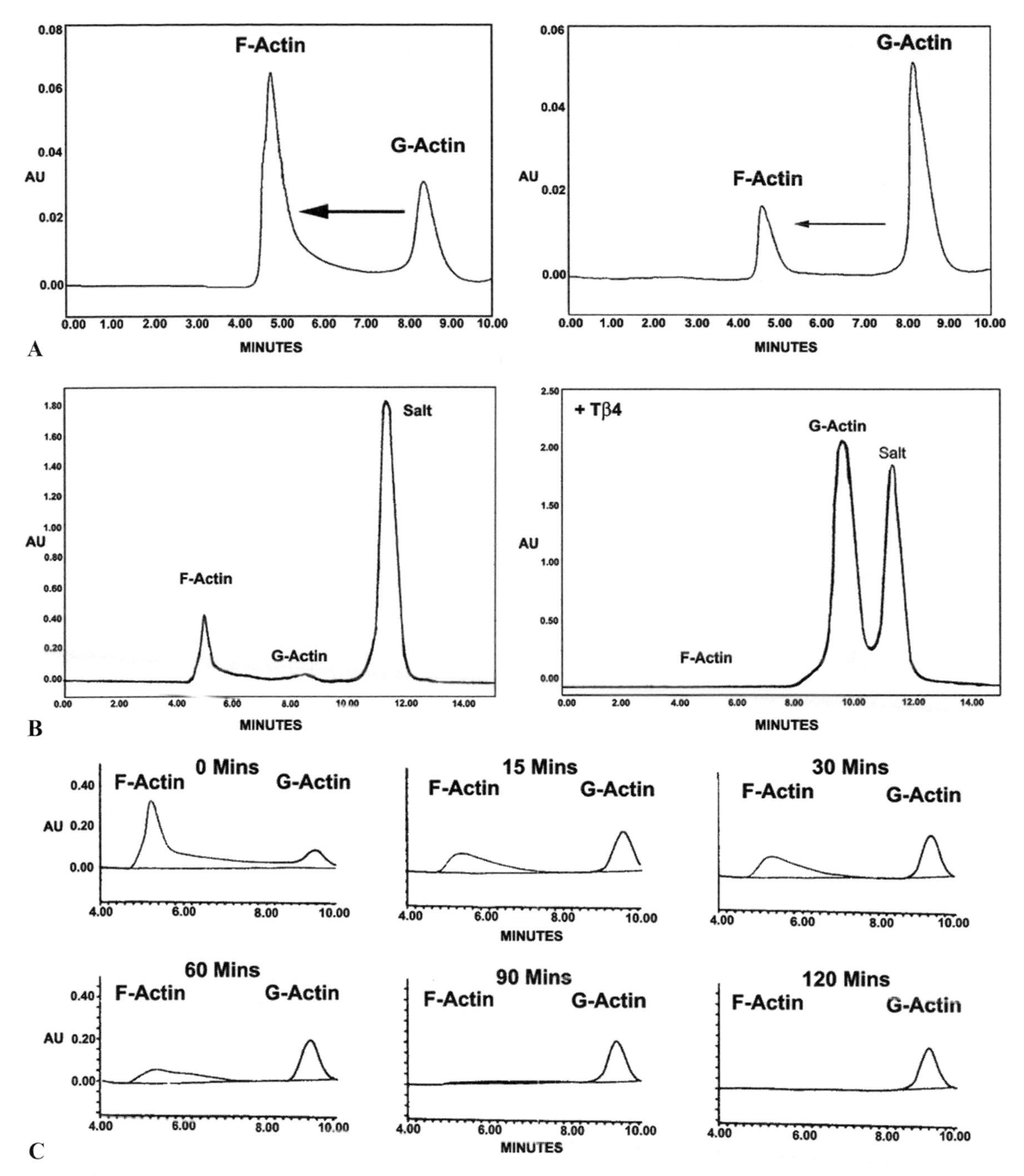

Figure 2. HPLC analysis of the effect of Tβ4 on actin polymerization and depolymerization. (A) 20 μM of G-actin was incubated with either 40 μM or 100 μM of Tβ4 for 60 min before HPLC. (B) 20 μM of F-actin was incubated in the presence or absence of Tβ4 for 60 min before HPLC. (C) F-actin incubated with Tβ4 at a molar ratio of 1:5 (actin/Tβ4) and was analyzed by HPLC at 0, 15, 30, 60, 90, and 120 minutes. Detection was at 280 nm, hence the absence of Tβ4 peaks. Data represent work in progress.

either alone or in combination for 60 min before HPLC separation. Tested individually, each mucolytic agent decreased the F-actin component of CF sputum in a dose-dependent manner (Fig. 4A). Of note, these mucolytics demonstrated cumulative activity as the combination of all three achieved the greatest decrease in F-actin of all test groups (Fig. 4B).

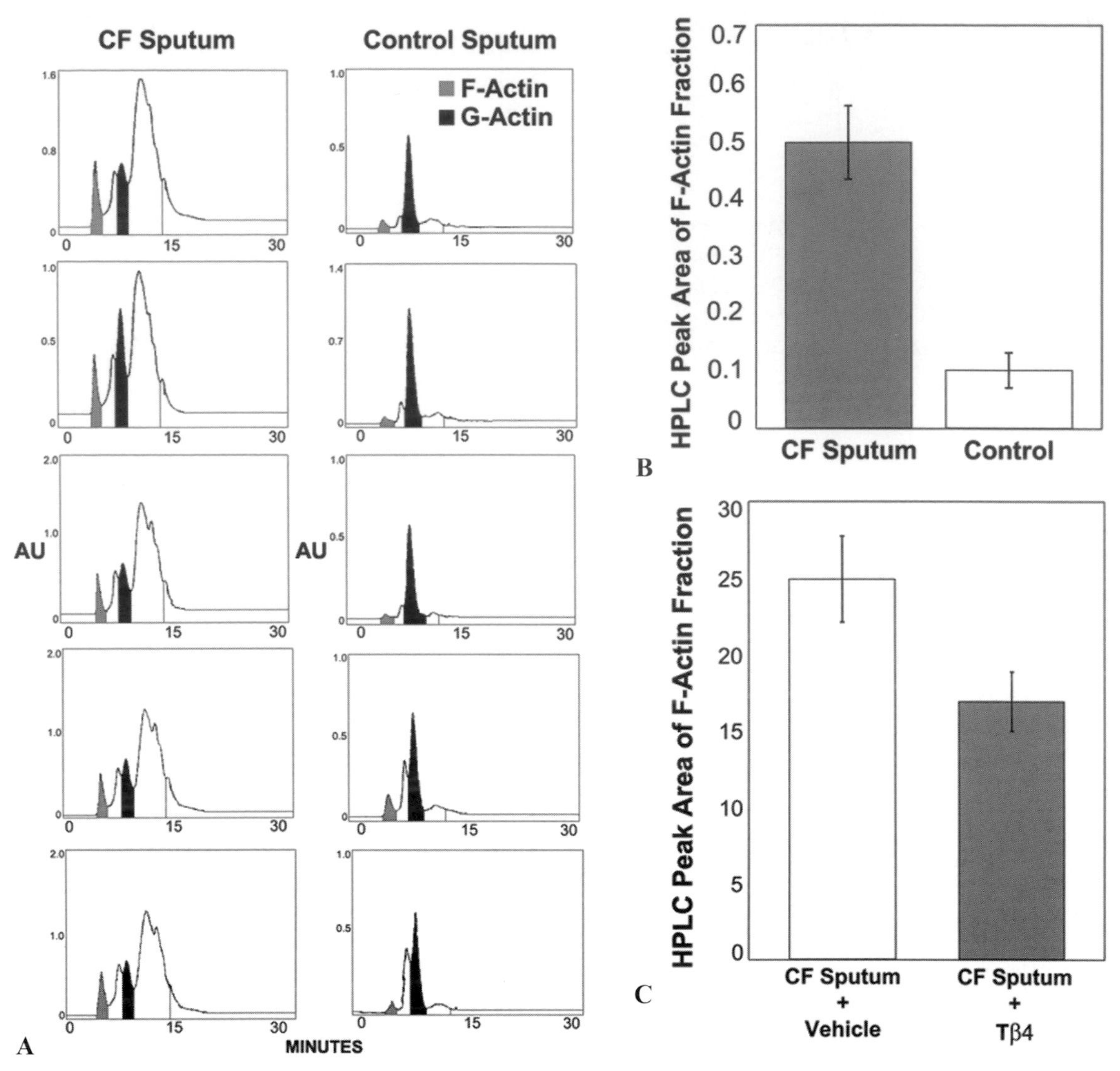

Figure 3. HPLC separation of CF sputum and normal control sputum. (A) Sputum samples of five CF patients and five healthy controls were solubilized in 0.1 M dithiothreitol-DDT at a ratio of 1:1 and centrifuged. 100 μL of the clear supernatants were used for HPLC analysis. (B) The ratio of F-actin to total actin in CF and control sputum samples were calculated using the HPLC peak area. (C) CF sputum samples were solubilized and centrifuged as above, and 150 μL of clear supernatants were incubated with either vehicle or 500 μg Tβ4 for 60 min before HPLC analysis. Peak areas of the F-actin fraction were calculated. Error bars are expressed as standard error of the mean. Data represent work in progress.

The additive nature of the effects of these molecules on decreasing F-actin content is likely a consequence of their differing yet related mechanisms of action. Both Tβ4 and DNase bind G-actin monomers at differing binding sites forming 1:1 complexes and functionally sequestering the monomers from assembly into F-actin filaments. DNase has the additional role of cleaving DNA polymers present in CF sputum, which has previously been postulated to increase the action of Tβ4 on actin by increasing the accessibility of actin-binding sites.[12] Further, the binding of G-actin by DNase significantly inhibits its action on DNA polymers, suggesting that concurrent use of Tβ4 serves to sequester G-actin and provide for more complete and uninhibited DNase activity. Gelsolin, on the other hand, functions primarily as an F-actin severing protein with the additional effect of capping the

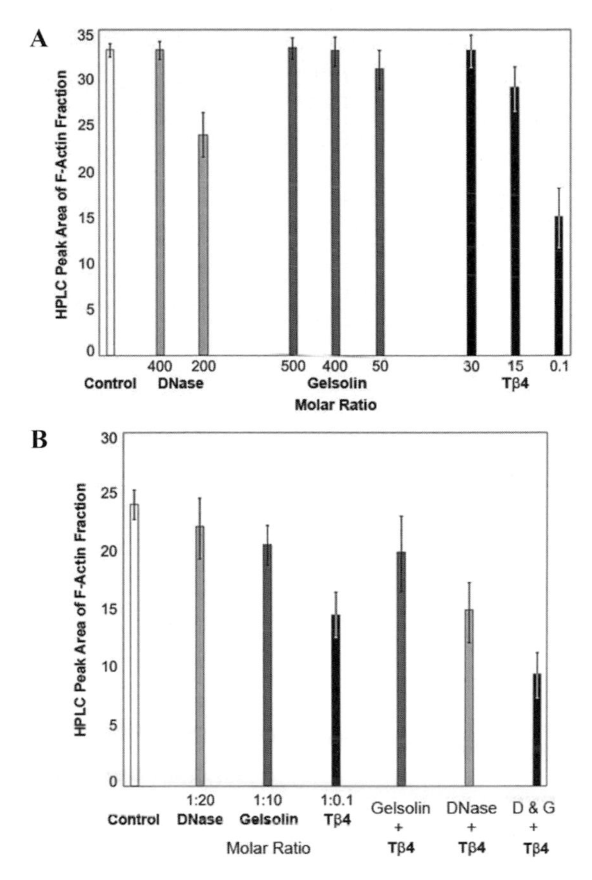

Figure 4. Dose–response effect of DNase, Gelsolin, and Tβ4 alone or in combination on F-actin depolymerization of CF sputum. CF sputum samples were incubated with different molar ratios of DNase, gelsolin, and Tβ4 (expressed as moles of test molecule per moles of F-actin) either (A) individually, or (B) in combination before HPLC separation. The HPLC peak area of the F-actin fraction was calculated. Error bars are expressed as standard error of the mean. Data represent work in progress.

growth end of F-actin filaments and thereby preventing elongation.

Fluorescence microscopy study

As a confirmatory step to corroborate the HPLC data presented, the effect of Tβ4 on the level of F-actin in CF sputum was examined using fluorescence microscopy. Sputum samples from the same patients were incubated with Tβ4 and then stained for F-actin and directly visualized by fluorescence microscopy. As has been shown in previous studies, Tβ4 significantly reduced the concentration and length of F-actin filaments in CF sputum (data not shown).

Discussion

The data summarized above indicate that this newly developed HPLC methodology is a rapid, reproducible, and quantitative technique to measure F-actin and G-actin in human sputum samples. Indeed, the favorable characteristics of HPLC, including its flexibility to perform a diversity of experiments from strict product quantitation to more elaborate time course and kinetic studies, as well as the relative ease to which such assays can be performed, should lend itself as a primary investigational modality in preclinical and *in vitro* studies of human sputum.

In patients with CF, large quantities of F-actin, as well as DNA, are typically found in the sputum and contribute significantly to the rigidity of the sputum.[7] It is generally accepted that these cellular constituents result from the necrosis of neutrophils and other white cell populations, as well as epithelial cells that line the airways, due to the chronic pulmonary infection and inflammation, which contributes to the eventual scarring of the membranes of the lungs. The preliminary observation that Tβ4 can reduce the rigidity of the sputum of CF patients by depolymerizing F-actin, as well as potentially reducing the scarring of the lungs by down-regulating a number of inflammatory chemokines and cytokines, suggests a potential clinical application in patients with CF.[12,14]

The principle of decreasing sputum cohesivity as a therapeutic approach to patients with cystic fibrosis has been confirmed by the successful application of dornase alfa in the clinical setting. Indeed, large-scale meta-analyses have demonstrated marked improvement in lung function as measured by spirometry in CF patients treated with dornase alfa at up to two years from initiation.[15] The data described above as well as previously published studies indicate an apparent additive effect of dornase alfa combined with Tβ4 in F-actin depolymerization and mucociliary transportability of CF sputum *in vitro*.[12] This cumulative benefit observed in these preclinical studies further suggests the possible use of Tβ4 in addition to dornase alfa in the clinical setting.

Conclusions

We have developed a new analytical methodology for measuring the content of F- and G-actin in the sputum of CF patients by HPLC. The assay is fast, quantitative, reproducible, and can easily be used to study the effect of new treatments on the structure and rigidity of CF sputum during drug development and clinical trials. Our studies also demonstrate that Tβ4 significantly increases the depolymerization of F-actin in CF sputum in a dose-dependent fashion. The results of this study provide evidence to support the potential use of Tβ4 in patients with CF.

Acknowledgments

The authors thank the Cystic Fibrosis Foundation for grant support that made this study possible, RegeneRx Biopharmaceutical Inc. for providing the Tβ4, and Robert J Fink, MD for his help in obtaining human sputum samples for the above studies.

Conflicts of interest

Authors M.B. and A.A.D. have no conflicts of interest. Author A.L.G. is Chairman of the Board and Chief Scientific Advisor of Regenerx Biopharmaceuticals Inc.

References

1. Lee, W.M. & R.M. Galibrith. 1992. The extracellular actin-scavenger system and actin toxicity. *N. Engl. J. Med.* **326, 20:** 335–340.
2. Stossel, T.P. 1993. On the crawling of animal cells. *Science* **260:** 106–109.
3. Pollard, T.D. & J.A. Cooper. 1986. Actin and actin-binding proteins. A critical evaluation of mechanisms and functions. *Annu. Rev. Biochem.* **55:** 987–1035.
4. Vasconcellos, C.A., P.G. Allen, M.E. Wohl, *et al.* 1994. Reduction of viscosity of cystic fibrosis sputum in vitro by gelsolin. *Science* **263:** 969–971
5. Potter, J., L.W. Mathews, J. Lemm, *et al.* 1960. *Am. I. Dis. Child.* **100:** 493–495.
6. Mornet, D. & K. Ue. 1984. Proteolysis and structure of skeletal muscle actin. *Proc. Natl. Acad. Sci. USA* **81:** 3680–3684.
7. Rubin, B.K. 2009. Mucous, phlegm, and sputum in cystic fibrosis. *Res. Care* **54:** 726–732.
8. Low, T.L.K., S.K. Hu & A.L. Goldstein. 1981. Complete amino acid sequence of bovine thymosin beta 4: a thymic hormone that induces terminal deoxynucleotidyl transferase activity in thymocyte populations. *Proc. Natl. Acad. Sci. USA* **78:** 1162–1166.
9. Sanders, M.C., A.L. Goldstein & Y.L. Wang. 1992. Thymosin beta 4 (Fx peptide) is a potent regulator of actin polymerization in living cells. *Proc. Natl. Acad. Sci. USA* **89:** 4678–4682.
10. Mora, C., C.A. Bauman, J.E. Paino, *et al.* 1997. Biodistribution of synthetic thymosin beta 4 in the serum, urine and major organs of mice. *Int. J. Immunopharmacol.* **19:** 1–8.
11. Badamchian, M., M. Fagarasan, R.L. Danner, *et al.* 2003. Thymosin beta(4) reduces lethality and down-regulates inflammatory mediators in endotoxin-induced septic shock. *Int. Immunopharmacol.* **3:** 1225–1233.
12. Rubin, B.K., A.P. Kater & A.L. Goldstein. 2006. Thymosin beta4 sequesters actin in cystic fibrosis sputum and decreases sputum cohesivity in vitro. *Chest* **130:** 433–1440.
13. Kater, A., M.O. Henke & B.K. Rubin. 2007. The role of DNA and actin polymers on the polymer structure and rheology of cystic fibrosis sputum and depolymerization by gelsolin or thymosin β4. *Ann. N.Y. Acad. Sci.* **1112:** 4–7.
14. Goldstein, A.L., E. Hannappel, G. Sosne & H.K. Kleinman. 2012. Thymosin β_4: a multi-functional regenerative peptide. Basic properties and clinical applications. *Expert Opin. Biol. Ther.* **12:** 37–51.
15. Jones, A.P. & C. Wallace. 2010. Dornase alfa for cystic fibrosis. *Cochrane Database Syst. Rev. Mar.* **17:** CD001127.

Ann. N.Y. Acad. Sci. ISSN 0077-8923

ANNALS OF THE NEW YORK ACADEMY OF SCIENCES
Issue: *Thymosins in Health and Disease*

The role of biologically active peptides in tissue repair using umbilical cord mesenchymal stem cells

Carlos Cabrera,[1] Gabriela Carriquiry,[1] Chiara Pierinelli,[1] Nancy Reinoso,[2] Javier Arias-Stella,[2] and Javier Paino[1]

[1]BioCell Peru Stem Cell Research Institute, Lima, Peru. [2]Arias-Stella Molecular Biology Laboratorios, Lima, Peru

Address for correspondence: Javier E. Paino, BioCell Peru Stem Cell Research Institute, Clinica San Felipe, G. Escobedo 676, Suíte 511, Lima 11, Peru. jpaino@neurocirugiaperu.com

The role of bioactive compounds in wound repair is critical. The preliminary work described herein includes the study of the effects of second degree burns in a Rex rabbit model and the action of human umbilical cord cells on the regulation and secretion of bioactive compounds. When applied on blood scaffolds as heterograft matrices, fibroblasts proliferate from these primary cultures and release biologically active peptides under tight control. Our work in progress indicates that mesenchymal stem cell (MSC)–mediated therapy provides better quality and more efficient burn reepithelialization of injured tissues by controlling the release of these peptides. Improvement of wound aesthetics is achieved in less time than without MSC-mediated therapy. Well-organized epidermal regeneration and overall better quality of reepithelialization, with no rejection, can be demonstrated consistently with periodic biopsies. Our studies indicate that MSCs have the capacity to produce, regulate, and deliver biologically active peptides that result in superior regeneration, compared with conventional treatments.

Keywords: umbilical cord cells; bioactive peptides; delivery systems

Introduction

Patients with dermal burns and lesions represent an important percentage of the total number of emergency cases that present in hospital emergency rooms (ER). To better help the thousands of patients seen in ER per year in both underdeveloped and developed countries, it is critical to understand and implement new and more efficient techniques in burn treatment. The beneficial role of stem cells, biologically active peptides, and biocompatible peptides on the mechanisms involved in tissue regeneration and repair is now well established.[1–3] However, attempts to make use of these cells or compounds directly on burn injuries to promote cell proliferation and healing have demonstrated that there are still difficulties in controlling their activity and determining the concentrations of active peptides needed for optimal wound healing.[4–7] Stem cells are readily available from umbilical cord blood from human donors or other mammals. They are well-known producers of very active biological compounds that serve the fetus in differentiating and regenerating tissue.[8]

Mechanism of tissue repair

Wound healing encompasses two separate processes: cell regeneration (or substitution for new healthy cells of the same linage) and replacement of normal skin tissue by fibroblasts and connective tissue. Both are based on the same cellular mechanisms: migration, proliferation, differentiation, and interaction with the cell matrix. An important issue to consider is that no new tissue can be layered and organized in the absence of healthy connective tissue and a basal membrane. Epithelial regeneration is an important clinical example of how new incoming cells intervene, not only in new cell formation but also in tissue connectivity, cell communication, and scaffolding.[7] These interactions are known to be promoted by precisely regulated bioactive proteins delivered by the surrounding tissue and new incoming cells.[8,9] Physiological processes involved in tissue regeneration include

doi: 10.1111/j.1749-6632.2012.06727.x

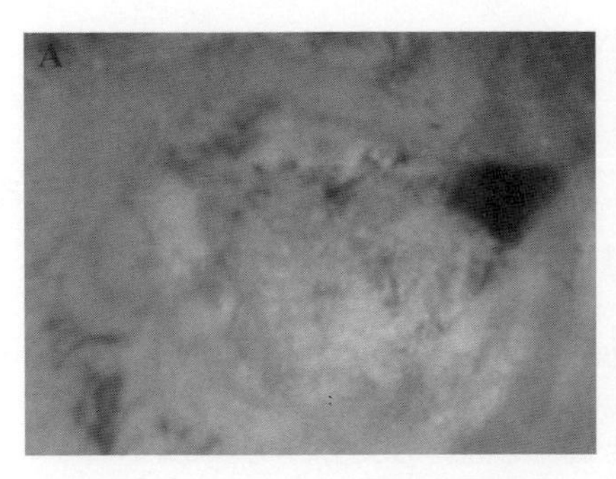

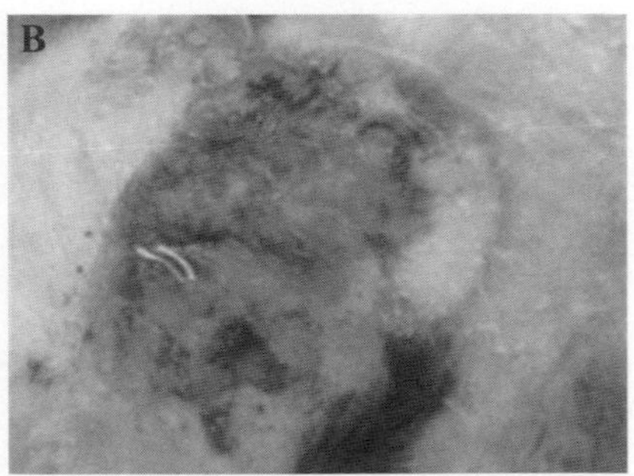

Figure 1. Postburn day 7 demonstrated a marked difference in appearance between (A) tissue undergoing umbilical cord MSC therapy and (B) tissue treated with a conventional adult dermal reticular compound. Note the well-organized and almost complete reepithelialization and wound contraction of the tissue treated with MSC therapy. These data represent work in progress.

inflammation, epithelialization, granulation tissue formation, neovascularization, wound retraction, and reorganization of the extracellular matrix.

Fetal tissue regeneration

Fetal tissue regeneration is significantly different than the reepithelialization seen in adult tissue. For example, it is much faster in the fetus because embryonic stem cells move from the base to the surface using actin fibers as virtual highways that allow cells to fill all empty spaces (as seen in Fig. 1A, work in progress). Studies demonstrate that healing in the fetus occurs with almost no scarring, most likely due to efficient cell migration and tight control of growth factor production and release.[10] Preliminary experiments with rats were carried out to confirm this hypothesis. Scarring was reduced by applying neutralizing antibodies against β transforming growth factors (TGF-β1, TGF-β2, and TGF-β3), whereas the control of TGF-β3 inhibits the transformation of the other two fractions, improving the quality of healing.[1,8]

Rex rabbit model for the study of the healing of burns

Much still remains to be understood about the mechanism of production and release of biologically active compounds by stem cells. To begin to address this issue, we have established a Rex rabbit model using the strict guidelines of the National Institutes of Health (NIH) for studies with animals. Concomitant diseases can influence healing; therefore, animals prepared for wound-healing studies must be examined to rule out underlying diseases and separated from the study if positive. Injuries such as second degree burns at 80 °C for 14 sec, with a round aluminium stamp and without additional pressure on a depilated dorsal skin, performed under sedation and with local anaesthesia, decreases animal stress, which may also influence healing. For that purpose, follow up analgesic treatment is preferable during the observation period. These rabbits have provided an excellent model for the study of burn healing.

The umbilical cord as a source of mesenchymal stem cells

Mesenchymal stem cells (MSCs) have the ability to proliferate and form extensive colonies of healthy newly differentiated fibroblasts. The newborn umbilical cord is a major, readily available source of potent MSCs.[11] These cells can be easily collected and cryopreserved. Modern tissue culture techniques have opened a vast horizon of mechanisms for cell growth control by closely regulating the production and release of bioactive proteins.[10] Also, a number of studies have demonstrated that there is a low risk of a graft-versus-host reaction with MSCs from newborn umbilical cords because the T lymphocytes in the umbilical cord tolerate the host HLA surface antigens very well. Conversely, there is high risk of histocompatibility rejection with adult T cells.

Stem cells withind the microenvironment of the wound appear to control the concentration and action of biologically active compounds, which, in turn, promote faster and more efficient healing. Cells do not need to be in direct contact with the injured tissue; their high capacity to produce biologically active peptides promotes regeneration even if these cells are removed from the area periodically. MSCs have the capacity to redirect their growth and differentiation via paracrine and autocrine growth regulation and programmed cell death.[1,3]

Umbilical cor blood is an excellent source of MSCs.[11] In our current work, we have used these cells and manipulated them *in vitro* with specific external growth factors to alter their characteristics. Colonies of cells capable of repairing damaged epithelium have been consistently obtained and are easily collected and cryopreserved. Sophisticated tissue culture methods allow manipulation and analysis of cell growth. Twenty MSC autocrine and paracrine stimulation and growth factor regulation can be studied closely. Keratinocytes and fibroblasts can also be cultured and studied for the purpose of developing cell treatments and biomaterials for tissue regeneration alternatives. Fibroblast cell lines derived from umbilical cord cells demonstrate activity in all stages of tissue regeneration. These cells create an appropriate environment for tissue regeneration above and beyond a "cell-for-a-cell" regeneration mechanism.

Stem cell–embedded scaffolds

Young adult Rex rabbits possess skin cells with healthy telomerases.[5] By covering half of acute second-degree burns with conventional protein-base gels, unenhanced results can be observed. Matrix-based cellular treatment can be compared side by side with the other half of the wound. Applying stem cells over the wound as a direct heterograft covering is not sufficient for control and regulation of healing compounds. To improve this initial deficient mechanism, we are working with a scaffold system from the same umbilical cord blood compounds manipulated in the lab. To apply cell therapy, scaffolds are embedded with transformed fibroblasts from primary cultures of MSC obtained from cord blood; the fibroblasts are seeded in a matrix and proliferate to confluence. The release of significant amounts of biologically active peptides is evident and measurable in these matrix-based cultures. Conventional adult cells and dermal reticular compounds can also be produced in culture dishes and used as secondary controls.

Overall, improvement in wound healing can be observed, primarily in the time and quality of the reepithelialization, compared with either MSC-embedded scaffold therapy or with conventional therapy. During the follow-up period, scaffolds can be changed up to three times, for faster epidermal regeneration and better quality of reepithelialization.

Differentiated fibroblasts in tissue regeneration

The best method to compare any conventional treatment with newly developed tissue regeneration techniques after burn injuries is to perform periodic biopsies of the healing tissue. In our ongoing preliminary study, biopsies are obtained at different points in time from the healing injuries to assess the appropriate use and efficiency of the different techniques used in the Rex rabbit model. No animals need to be sacrificed for this, and they can be returned to the animal facility after their wounds have healed, for further studies.

The use of cultured human umbilical cord cells has resulted in an overall better quality of reepithelilization without rejection. Cells obtained from voluntary umbilical cord donors can be consistently turned into MSCs. Cells are characterized using flow cytometry for the MSC markers CD73, CDw90, CD105, and CD166, and the hematopoietic

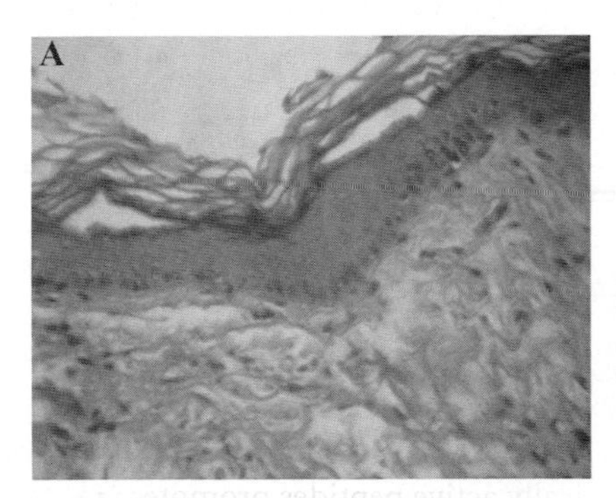

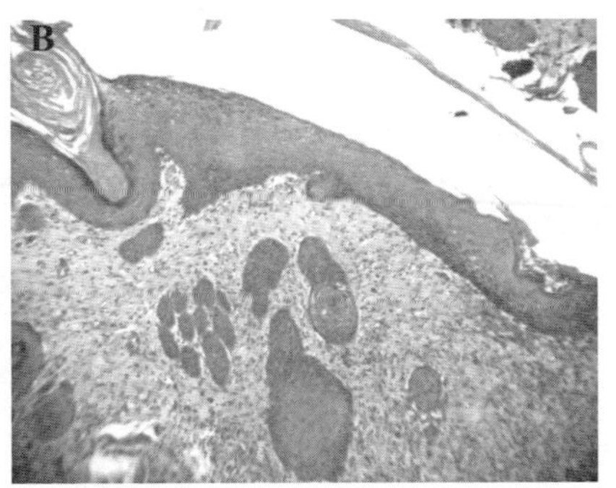

Figure 2. Biopsy at day 12 demonstrates (A) a well-organized epithelium in umbilical cord cell-treated tissue, with a uniform and continuous basal membrane, stratified epithelium, and normal thickness corneal layer. (B) Shows a disorganized hyperplastic dermis, discontinuous basal membrane, and disorganized epidermis layers. These data represent work in progress.

markers CD34, CD45, CD117, and CD123 (for isotype control, IgG-FITC and IgG-PE can be used). After several passages, strong, newly transformed fibroblasts are allowed to grow to 100% confluence and are later characterized by vimentin immunohistochemistry. Fibroblast-embedded fibrin matrices can be prepared with plasma from the residual umbilical cord blood.[8,12]

To provide this preparation with a steady scaffold, basal structures are prepared using a gel obtained from sodium carboxymethylcellulose, pectin, and propyleneglycol (DuoDERM® extra thin).[13–15] For control groups, a collagen-based adult dermal reticular compound with bovine collagen and chondroitin 6 sulfate, with a silicone base external layer, is conventionally used.[2,16] This compound is used in mainstream treatment in major burn units with acceptable results.[4,17]

Fibroblasts are crucial in tissue repair. These cells are capable of autocrine and paracrine regulation. Their differentiation, growth, and interaction is modulated with precision;[3,18] growth factor receptors act inside the cell via tyrosine kinases. Fibroblasts work on tissue repair by synthesizing fibers, including reticular, collagen, and elastin fibers. During the healing process, fibroblasts migrate and proliferate to form the scaffold in which epithelial cells proliferate and regenerate the skin in an organized progression (as seen in Fig. 2A).[13] In contrast to well-organized healing , fibroblasts are responsible for replacing all necrotic layers when the burn is too severe (as seen in Fig. 2B). Fibroblasts produce and regulate the concentration of bioactive proteins in the injured area. These cells use bioactive peptides, such as PDGF, to promote chemotactic effects on macrophages and ECGF for neovascularization. It has been demonstrated that these peptides are important during every step of skin repair, macrophage migration, granulation tissue formation, and extracellular matrix organization.[7] During a period of approximately one to three days, active angiogenesis and remodeling occur. Later, a process called wound retraction, in which fibroblasts acquire a differentiated phenotype and start a process of tissue compaction, occurs.

Conclusions

Much remains to be studied about the influence of MSCs in the environment where they are placed. We believe that these cells play an important role in the release of bioactive peptides, their delivery, and regulation of the concentration of these molecules in the local microenvironment, and their activity—more so than just a quantitative replacement of missing tissue. We aim to demonstrate in our ongoing studies that stem cells are capable of controlling production and delivery of peptides, such as thymosin β4, which are known key players in tissue regeneration in the fetus and newborn in the repair mechanisms associated with wound healing. We believe one of the most important mechanisms of action of these cells is the tight control they have over the release and activity of such peptides in the microenvironment. Therefore, any attempt on implementing innovative treatments using stem cells should take into consideration this very powerful tool for delivery and regulation of bioactive peptides, as opposed to only applying peptides, growth factors, or cells randomly within the wound bed.

Conflicts of interest

The authors declare no conflicts of interest.

References

1. Abkowitz, J. 2002. Can human haematopoietic stem cells become skin, gut, or liver cells? *N. Engl. J. Med.* **346:** 770–772.
2. Badylak, S.F. 2004. Xenogeneic extracellular Matrix as scaffold for tissue reconstruction. *Transplant. Immunol.* **12:** 367–377.
3. Bruder, S. *et al.* 1998. Growth kinetics, self-renewal, and the osteogenic potential of purified human mesenchymal stem cells during extensive subcultivation and following cryopreservation. *J. Cell. Biochem.* **62:** 278–294.
4. Kaihara, S. *et al.* 1999. Tissue engineering towards new solutions for transplantation and reconstructive surgery. *Arq. Surg.* **134:** 1184–1188.
5. Knabl, J. *et al.* 1999. Controlled partial skin thickness burns: animal model for studies of burn wound progression. *Burns* **25:** 229–235.
6. Rheinwald, J.G. & Green, H. 1975. Serial cultivation of strains of human epidermal keratinocytes: the formation of keratinising colonies from single cells. *Cell* **6:** 331–334.
7. Singer, A. *et al.* 1999. Mechanism of disease: cutaneous wound healing. *N. Engl. J. Med.* **341:** 738–746.
8. Campagnoli, C. *et al.* 2001. Identification of mesenchymal/progenitor cells in human first trimester fetal blood, liver, and bone marrow. *Blood* **98:** 2396–2402.
9. Korbling, M. *et al.* 2003. Medical progress: adult stem cell for tissue repair; a new therapeutic concept? *N. Engl. J. Med.* **349:** 570–582.
10. Mareschi, K. *et al.* 2001. Isolation of human mesenchymal stem cells: bone marrow versus umbilical cord blood. *Haematologica* **86:** 1099–1100.

11. Romanov, Y. *et al.* 2003. Searching for alternative sources of postnatal human mesenchymal stem cells (MSC): candidate MSC-like cells from umbilical cord. *Stemcells* **21:** 105–110.
12. Majumdar, M.K. *et al.* 1998. Phenotypical and functional comparison of cultures of marrow derived MSC and stromal cells. *J. Cell Phys.* **176:** 57–66.
13. Atala, A. *et al.* 1992. Formation of urothelia structures in vivo from dissociated cells attached to biodegradable polymer scaffolds in vitro. *J. Urol.* **148:** 658–662.
14. Fijiyama, C. *et al.* 1995. Reconstruction of the urinary bladder mucosa in three-dimensional collagen gel culture: fibroblast extracellular matrix interactions on the differentiation of transitional epithelial cells. *J. Urol.* **153:** 2060–2067.
15. Ozeki, M. & Y. Tabata. 2005. In vivo degradability of hydrogels prepared from different gelatins by various crosslinking methods. *J. Biomater Sci. Polym. Ed.* **16**(5)**:** 549–561.
16. Buckley, C. & K. O'Kelly. 2004. *Regular Scaffold Fabrication Techniques for Investigation in Tissue Engineering*. Centre for Bioengineering Department of Mechanical and Manufacturing Engineering, Trinity College. Dublin, Ireland. 147–166.
17. Liang, D. *et al.* 2007. Functional electrosporum nanofibrous scaffolds for biomedical applications. *Adv. Drug Deliv. Rev.* **59:** 1392–1412, 0169–0409.
18. Pittenger, M.F. *et al.* 1999. Multilinage potential of adult human mesenchymal stem cells (MSC). *Science.* **284:** 143–147.
19. Gutierrez-Rodriguez, M. *et al.* 2000. Characterization of the adherent cells developer in Dexter type long-term cultures from human umbilical cord blood. *Stemcell* **18:** 46–52.
20. Takahashi, K. *et al.* 2006. Induction of pluripotent stem cells from mouse embryonic and adult fibroblast cultures by defined factors. *Cell* **126:** 663–676.

Ann. N.Y. Acad. Sci. ISSN 0077-8923

ANNALS OF THE NEW YORK ACADEMY OF SCIENCES
Issue: *Thymosins in Health and Disease*

Identification of interaction partners of β-thymosins: application of thymosin β4 labeled by transglutaminase

Christine App,[1] Jana Knop,[1] Hans Georg Mannherz,[2] and Ewald Hannappel[1]

[1]Institute of Biochemistry, Friedrich Alexander University, Erlangen, Germany. [2]Department of Anatomy and Molecular Embryology, Ruhr University, Bochum and Max-Planck-Institute of Molecular Physiology, Dortmund, Germany.

Address for correspondence: Christine App, Institute of Biochemistry, Friedrich-Alexander-University, Fahrstrasse 17, 91054 Erlangen, Germany. thymosin@biochem.uni-erlangen.de

In this review, we identify potential interaction partners of the β-thymosin family. The proteins of this family are highly conserved peptides in mammals and yet only one intracellular (G-actin) and one cell-surface protein (β subunit of F_1–F_0 ATP synthase) were identified as interaction partners of thymosin β4. Cross-linking experiments may be a possible approach to discover additional proteins that interact with the β-thymosin family. It has previously been shown that thymosin β4 can be labeled at its glutaminyl residues with various cadaverines using tissue transglutaminase. Here, we illuminate recent results and give an outlook on upcoming work in the field.

Keywords: thymosin β4; transglutaminase; label; cadaverine; putrescine; actin; cross-link

Introduction

In living cells, proteins interact with various biomolecules such as metal ions, lipids, sugars, and other proteins. Specific interactions are critical for biological function and regulation. *In vitro* protein analyses do not always sufficiently elucidate potential interaction partners because some of the interactions are likely to be destroyed through the analysis. Unfortunately, it is not currently possible for scientists to examine all of the ongoing processes at the cellular level in which the protein of interest is involved.

In the last few years, a common method was to isolate the protein of interest and study its behavior *in vitro*. This approach, however, is very destructive and can lead to false negative results because weak interactions may easily be rated as irrelevant or lost though they might be important at physiological conditions. On the other hand, this experimental procedure can also lead to false positive results in which molecules appear to interact with partners that are not present in the same intracellular compartment *in vivo*. Despite the fact that the dissociation constant of the complex of thymosin β4 with actin is only in the micromolar range,[1] thymosin β4 is the main actin-sequestering molecule in the mammalian cell.[2–5] The reversibility of actin sequestration is indeed dependent on the relatively low affinity of β-thymosins for G-actin; thus, this process is readily reversible and actin can be intracellularly repolymerized after cell stimulation.[5]

A powerful method to identify interactions is to cross-link the protein of interest to the interacting partner. A possible way of doing this is to conduct photoaffinity labeling experiments.[6,7] In the last few years, several new techniques were developed that allow photoreactive crosslinkers to be incorporated into biological molecules in a site- and residue-specific manner. The advantage of these crosslinkers is that their functional groups can be activated by light in a time- and space-dependent manner. Upon photoactivation, this type of crosslinker is converted to highly reactive species that can chemically react with neighboring structures. This enables scientists to obtain a "snapshot" of the cellular environment of a molecule.

To work in a cellular environment, photocrosslinkers must meet additional requirements. The unactivated form of the crosslinker must not change the cellular behavior of the investigated protein and should be stable enough to remain

doi: 10.1111/j.1749-6632.2012.06658.x

unactivated until irradiated with the appropriate wavelength of light.[8] The four most commonly used photocrosslinking groups are benzophenone, aryl azide, trifluormethyl phenyl diazirine, and alkyl diazirine. Additional desirable characteristics are small size, long-excitation wavelength, and the ability to capture specific binding events with high cross-linking efficiency.[8] To minimize the disturbance of the interaction a small size of the crosslinker is preferable.

Thymosin β4 is the most prominent member of the β-thymosin family. It is considered to be a moonlighting polypeptide with multiple biological functions.[9] Angiogenic effects of thymosin β4 have been shown *in vitro* and *in vivo*.[10–12] Thymosin β4 promotes endothelial cell migration *in vitro*[11,13] and accelerates wound healing in rats[12] and in the human eye.[13,14] A study in human platelets has identified thymosin β4 as one of seven thrombin-releasable antimicrobial peptides.[15]

Despite these multiple functions of thymosin β4, interactions with only two proteins have been reported. One of the interacting proteins is intracellular. In 1991, Safer *et al.* reported that the actin-sequestering protein Fx and thymosin β4 are indistinguishable.[16] The other interacting protein is the F_1–F_0 ATP-synthase, which is located on the surface of HUVEC cells.[17] It was identified by a pull-down experiment with an N-terminal biotinylated thymosin β4 using a membrane fraction of HUVECs and characterized by MALDI-TOF analysis. Because of these data, we searched for possible cross-linking approaches to identify further interaction partners of thymosin β4.

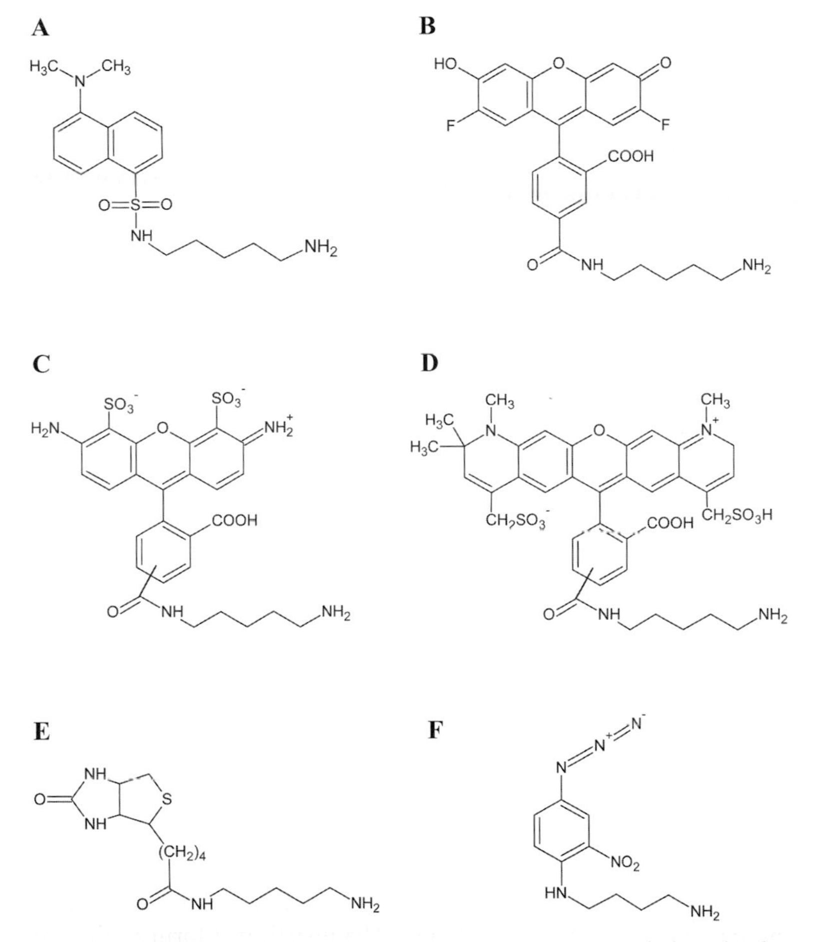

Figure 1. Structures of cadaverine- and putrescine-based reagents. All of these reagents can be bound to thymosin β4 using tissue transglutaminase. (A) Dansyl cadaverine; (B) Oregon Green cadaverine; (C) Alexa488 cadaverine; (D) Alexa594 cadaverine; (E) Biotin cadaverine; (F) *N*-(4-azido-2-nitrophenyl)-putrescine (ANP).

Thymosin β4 and transglutaminase

Tissue transglutaminases catalyze a calcium-dependent acyl transfer reaction between the γ-carboxamide group of a peptide-bound glutamine residue and the primary amino group of either a peptide-bound lysine or a polyamine.[18] The bond generated by tissue transglutaminase is covalent, stable, and resistant to proteolysis. Transglutaminases have a broad specificity for primary amine substrates, whereas the number of proteins that serve as glutaminyl substrates is highly restricted.[19]

It has been shown that thymosin β4 can serve as a specific glutaminyl substrate of guinea pig transglutaminase *in vitro.*[19] The three glutaminyl residues of thymosin β4 can be cross-linked by tissue transglutaminase to cadaverines and ε-amino groups of several proteins.[20] The reaction of amino acid residue Q_{23} and Q_{36} to thymosin β4 is fast, whereas Q_{39} reacts only very slowly. Q_{23} and Q_{36} are conserved in all β-thymosins except for thymosin β15 in which Q_{23} is replaced by glutamate.[21]

Transglutaminase can be used to attach various cadaverines to thymosin β4. In the past we used Dansyl cadaverine,[19,22] Biotin cadaverine, Oregon Green cadaverine,[23,24] Alexa488 cadaverine, and Alexa594 cadaverine to modify thymosin β4. These fluorescent derivatives of β-thymosins are useful tools to elucidate its intracellular distribution in cells that can be studied by microinjection.[20] We demonstrated that thymosin β4 is present not only in the cytoplasm but also in the nucleus of cells.[23] Lately, there has been increasing evidence for the presence of cytoskeletal proteins in the nucleus, such as actin itself,[25,26] actin-related proteins (Arps),[27–29] and a number of different actin-binding proteins.[30–33] Thymosin β4 in the nucleus might be involved in actin sequestering, especially because no nuclear F-actin can be detected by fluorescently labeled phalloidin.[23] Although the relevance of nuclear thymosin β4 is not yet completely understood, there is evidence that cytoskeletal proteins are involved in activities ranging from nuclear assembly to DNA replication and transcription.[23,34–36]

The structures of all cadaverines used up to now for modifying thymosin β4 are shown in Figure 1. Recently, thymosin β4 has been labeled with *N*-(4-azido-2-nitrophenyl)-putrescine (ANP, Fig. 1F), a heterobifunctional photocrosslinker.[37] This putrescine derivative has the advantage over the fluorescent cadaverines that it can be cross-linked by UV-light to the proteins in close vicinity. The distance spanned by ANP is between 11.1 and 12.5 Å.[37] The putrescine moiety of ANP serves as a substrate for tissue transglutaminase, which makes it possible to attach ANP to glutaminyl residues of thymosin β4. The photoreactive group of ANP is an azido group. Aryl azides cross-link through a reactive species produced by loss of nitrogen and concomitant formation of a

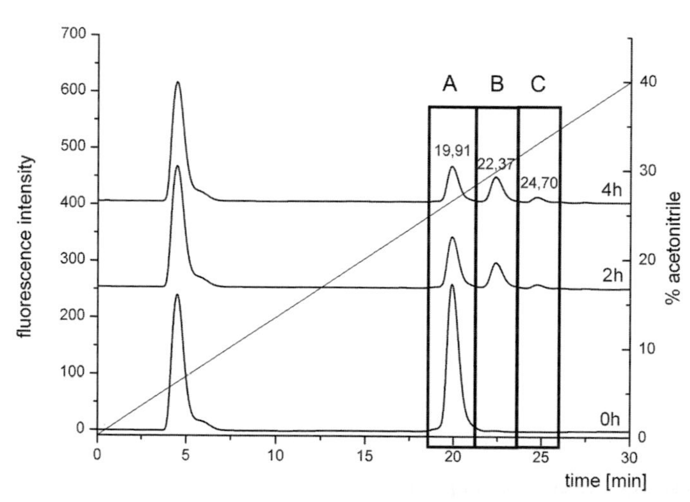

Figure 2. Time course of labeling of thymosin β4 using transglutaminase. Aliquots were taken from the reaction mixture after incubation for 0, 2, and 4 hours. Gradient was linear in 30 min to 40% acetonitrile in 0.1% TFA. Peptides were isolated by preparative RP-HPLC and subjected to mass spectrometry.

nitrene.[23] Once formed, nitrenes can insert into neighboring C–H and heteroatom–H bonds. This broad reactivity allows aryl azides to react with a variety of biomolecules. If a suitable interacting partner for reaction is not available, the nitrene will typically persist for about 0.1 ms before converting to a less reactive but more stable ketenimine.[8] The specificity of the photoactivated reaction of the ANP is high and limited by diffusion.[37]

The typical time course of a labeling reaction of thymosin β4 is exemplarily shown in Figure 2. Thymosin β4 was incubated with an eight-fold molar excess of ANP, and the reaction was started by the addition of guinea pig transglutaminase. RP-HPLC analyses done directly after addition of the enzyme and after two and four hours are depicted (Fig. 2). The beginning of the reaction analysis by RP-HPLC showed a large peak for thymosin β4 (peak A). After two and four hours the amount of thymosin β4 was diminished while two new peaks (B and C) appeared. Several fractions were isolated using preparative HPLC and characterized by mass spectroscopy using MALDI-TOF. The data obtained through MALDI-TOF analysis showed that there was still unreacted thymosin β4 (peak A) present. The observed data corresponding to peak B showed a peptide of the expected mass of monolabeled thymosin $β_4$, while peak C is bislabeled thymosin β4.

Techniques to identify labeled glutaminyl residues

In previous experiments using Dansyl cadaverine as a labeling reagent, it has been shown that two of the three glutaminyl residues of thymosin β4 are more likely to be modified, namely Q_{23} and Q_{36}.[19] To determine which glutaminyl residues are labeled a proteolytic digest of the labeled peptides is a possibility. A digest with AsnC-endoproteinase yields two fragments of thymosin β4.[23] Both fragments hold one of the reactive glutaminyl residues. Another way to verify the modified glutaminyl residue would be a tryptic digest of thymosin β4, which leads to a tryptic fingerprint.[38] The glutaminyl residues reside in different fragments formed by the cleavage (Fig. 3A). As an example of this method, the tryptic fingerprint of thymosin β4 and its ANP derivatives is depicted in Figure 3B. The tryptic fingerprint of thymosin β4 yielded nine fragments, T2 contained Q_{23}, while T6 contained Q_{36}. The digest of the

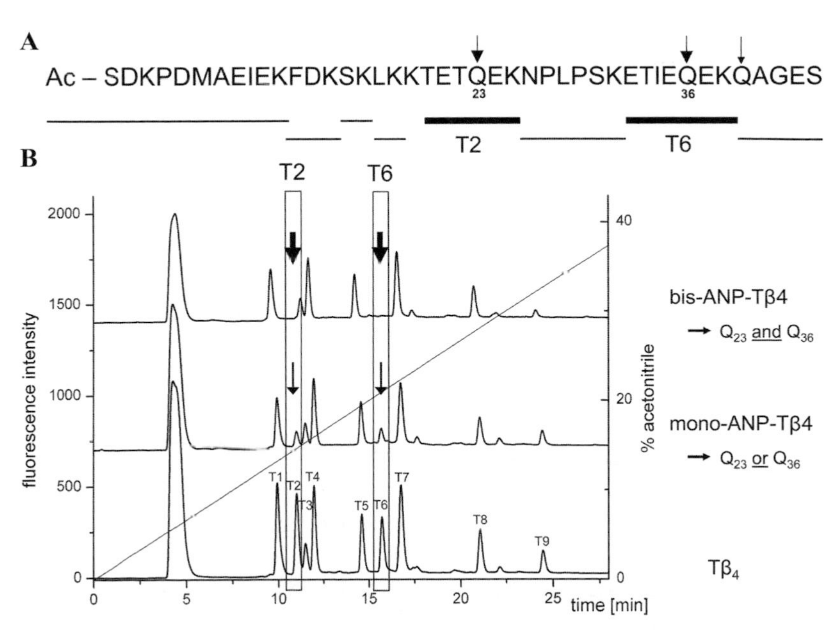

Figure 3. (A) Amino acid sequence of thymosin β4. The three arrows point to the glutaminyl residues serving as potential glutaminyl substrates for tissue transglutaminase. Tryptic fragments of thymosin β4 are shown by lines. The tryptic fragments T2 and T6 are depicted as bold lines containing the highly reactive Q_{23} or Q_{36}. (B) RP-HPLC analysis of tryptic fragments generated from thymosin β4 and ANP-labeled peptides. The fragments of interest are marked with a box. The extent of the decrease of T2 and T6 is indicated by the thickness of the arrows within the box.

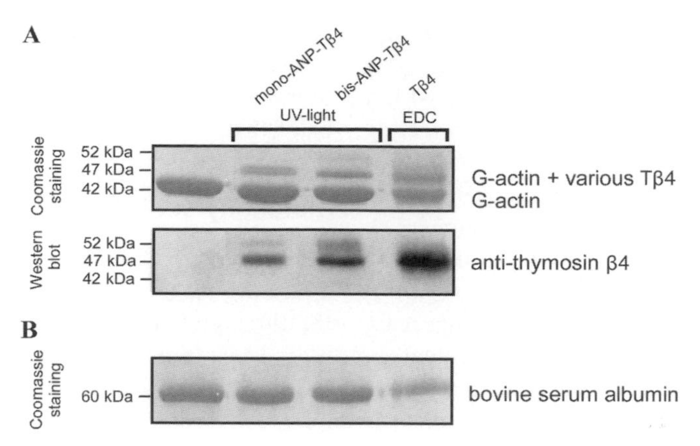

Figure 4. Cross-linking of ANP-labeled derivatives or thymosin β4 to G-actin. (A) G-actin was irradiated in the presence of ANP-thymosin β4 derivatives or incubated with thymosin β4 and 1-ethyl-3-(3-dimethylaminopropyl)carbodiimide (EDC). Detection of proteins was performed by Coomassie blue staining (A, upper half) or Western blot analysis (A, lower half). (B) As a control, G-actin was replaced by bovine serum albumin; all other conditions are as in panel A.

monolabeled derivative resulted in a decrease of tryptic fragments T2 and T6. This led to the conclusion that the monolabeled thymosin β4 is either modified at Q_{23} or Q_{36}. The tryptic fingerprint of the digested bislabeled derivative showed no detectable peaks for fragments T2 and T6. Therefore, this derivative is labeled at both glutaminyl residues (Q_{23} and Q_{36}).

Biological activity of thymosin β4 labeled by tissue transglutaminase

In previous experiments using cadaverines to modify thymosin β4 we were able to show that all these derivatizations surprisingly do not interfere with the ability of thymosin β4 to sequester G-actin.[19,23] Therefore, it has been concluded that Q_{23} and Q_{36}, though conserved in mammalian thymosin β4,[21] are not involved in G-actin sequestering. This has been verified by determination of the K_d-values of the complex of G-actin with Oregon Green cadaverine–labeled thymosin β4, which did not differ from the values determined for the complex between G-actin and unlabeled thymosin β4.[23]

Another aspect studied was the capacity of the derivatized β-thymosins to inhibit salt-induced G-actin polymerization using falling ball viscosimetry. The viscosity of polymerized actin of a suitable concentration was determined by viscosimetry. At equimolar concentrations of G-actin and thymosin $β_4$, the value decreased to the viscosity of the buffer solution and represented complete inhibition of polymerization. Equal amounts of thymosin β4 derivatives, for example mono-ANP-thymosin β4, also inhibited the polymerization to a similar extent. Therefore, the modification of the glutaminyl residues of thymosin β4 does not interfere with the ability to inhibit G-actin polymerization.

Thymosin β4 is a G-actin sequestering peptide and can be cross-linked to G-actin by 1-ethyl-3-(3-dimethylaminopropyl)carbodiimide (EDC), a zero-length crosslinker.[16] EDC generates isopeptide bonds between carboxyl and amino groups only in very close vicinity. The steric requirements of the reaction are stringent. It has been shown that labeled thymosin β4 can be cross-linked to G-actin using EDC.[19,23] Crosslinkers bridging longer distances would be a helpful tool to identify additional proteins interacting with thymosin β4. ANP is a photoactivable crosslinker. To cross-link the ANP-labeled thymosin β4 derivatives to G-actin, both proteins were incubated at an equimolar ratio and then cross-linked using UV light. Thymosin β4 cross-linked to actin by EDC served as a positive control. The samples were separated by SDS-PAGE and the gel was stained with Coomassie blue (Fig. 4A, upper panel). It was possible to cross-link all ANP derivatives of thymosin β4 to actin. Even some double cross-links were observed (actin bound to two molecules of derivatized thymosin β4). In a Western blot analysis, the cross-linked products were still recognized by a thymosin β4-antibody (Fig. 4A, lower panel). The UV-induced cross-linking reaction of the ANP derivatives of thymosin β4 to G-actin is specific because no cross-linking products were observed when G-actin is replaced by bovine serum albumin in the irradiated mixture.

Conclusion

It is possible to bind cadaverine and putrescine derivatives covalently to thymosin β4 using guinea pig transglutaminase. Out of the three glutaminyl residues of thymosin β4, Q_{23} and Q_{36} are more likely to be modified. It is possible to isolate different derivatives, both monolabeled and bislabeled. Normally the monolabeled fraction is modified at either one of the two reactive glutaminyl residues (Q_{23} or Q_{36}), whereas the bislabeled-thymosin β4 is labeled at both reactive residues (Q_{23} and Q_{36}). All derivatives still inhibit salt-induced polymerization of G-actin. It is possible to cross-link the derivatives specifically to G-actin using EDC or UV-light.

UV-light activable derivatives of thymosin β4 will facilitate the identification of additional interacting partners in future.

Acknowledgments

The authors would like to thank György Heygi (Budapest, Hungary) for the ANP. We would like to express our appreciation to Doris Jaegers for excellent technical assistance, and H.G.M. thanks the Deutsche Forschungsgemeinschaft (DFG) for financial support.

Conflicts of interest

The authors declare no conflicts of interest.

References

1. Huff, T. *et al.* 2001. β-Thymosins, small acidic peptides with multiple functions. *Int. J. Biochem. Cell Biol.* **33:** 205–220.
2. Hannappel, E., S. Davoust & B.L. Horecker. 1982. Thymosins $β_8$ and $β_9$: two new peptides isolated from calf thymus homologous to thymosin $β_4$. *Proc. Natl. Acad. Sci. USA* **79:** 1708–1711.
3. Hannappel, E. & W. Leibold. 1985. Biosynthesis rates and content of thymosin $β_4$ in cell lines. *Arch. Biochem. Biophys.* **240:** 236–241.
4. Xu, G.J. *et al.* 1982. Synthesis of thymosin $β_4$ by peritoneal macrophages and adherent spleen cells. *Proc. Natl. Acad. Sci. USA* **79:** 4006–4009.
5. Mannherz, H.G. & E. Hannappel. 2009. The β-thymosins: intracellular and extracellular activities of a versatile actin binding protein family. *Cell Motil. Cytoskeleton* **66:** 839–851.
6. Brunner, J. 1993. New photolabeling and crosslinking methods. *Annu. Rev. Biochem.* **62:** 483–514.
7. Singh, A., E.R. Thornton & F.H. Westheimer. 1962. The photolysis of diazoacetylchymotrypsin. *J. Biol. Chem.* **237:** 3006–3008.
8. Tanaka, Y., M.R. Bond & J.J. Kohler. 2008. Photocrosslinkers illuminate interactions in living cells. *Mol Biosyst.* **4:** 473–480.
9. Goldstein, A.L., E. Hannappel & H.K. Kleinman. 2005. Thymosin $β_4$: actin sequestering protein moonlights to repair injured tissues. *Trends Mol. Med.* **11:** 421–429.
10. Grant, D.S. *et al.* 1995. Matrigel induces thymosin $β_4$ gene in differentiating endothelial cells. *J. Cell Sci.* **108**(Pt 12): 3685–3694.
11. Malinda, K.M., A.L. Goldstein & H.K. Kleinman. 1997. Thymosin $β_4$ stimulates directional migration of human umbilical vein endothelial cells. *FASEB J.* **11:** 474–481.
12. Malinda, K.M. *et al.* 1999. Thymosin $β_4$ accelerates wound healing. *J. Invest. Dermatol.* **113:** 364–368.
13. Sosne, G. *et al.* 2001. Thymosin $β_4$ promotes corneal wound healing and modulates inflammatory mediators in vivo. *Exp. Eye Res.* **72:** 605–608.
14. Sosne, G. *et al.* 2002. Thymosin $β_4$ promotes corneal wound healing and decreases inflammation in vivo following alkali injury. *Exp. Eye Res.* **74:** 293–299.
15. Tang, Y.Q., M.R. Yeaman & M.E. Selsted. 2002. Antimicrobial peptides from human platelets. *Infect. Immun.* **70:** 6524–6533.
16. Safer, D., M. Elzinga & V.T. Nachmias. 1991. Thymosin $β_4$ and Fx, an actin-sequestering peptide, are indistinguishable. *J. Biol. Chem.* **266:** 4029–4032.
17. Freeman, K.W., B.R. Bowman & B.R. Zetter. 2011. Regenerative protein thymosin $β_4$ is a novel regulator of purinergic signaling. *FASEB J.* **25:** 907–915.
18. Greenberg, C.S., P.J. Birckbichler & R.H. Rice. 1991. Transglutaminases: multifunctional cross-linking enzymes that stabilize tissues. *FASEB J.* **5:** 3071–3077.
19. Huff, T. *et al.* 1999. Thymosin $β_4$ serves as a glutaminyl substrate of transglutaminase. Labeling with fluorescent dansylcadaverine does not abolish interaction with G-actin. *FEBS Lett.* **464:** 14–20.
20. Hannappel, E. 2007. β-Thymosins. *Ann. N. Y. Acad. Sci.* **1112:** 21–37.
21. Goldstein, A.L. *et al.* 2012. Thymosin $β_4$: a multi-functional regenerative peptide. Basic properties and clinical applications. *Expert Opin. Biol. Ther.* **12:** 37–51.
22. Ballweber, E. *et al.* 2002. Polymerisation of chemically cross-linked actin:thymosin $β_4$ complex to filamentous actin: alteration in helical parameters and visualisation of thymosin $β_4$ binding on F-actin. *J. Mol. Biol.* **315:** 613–625.
23. Huff, T. *et al.* 2004. Nuclear localization of the G-actin sequestering peptide thymosin $β_4$. *J. Cell Sci.* **117:** 5333–5343.
24. Zoubek, R.E. & E. Hannappel. 2007. Subcellular distribution of thymosin $β_4$. *Ann. N. Y. Acad. Sci.* **1112:** 442–450.
25. Gonsior, S.M. *et al.* 1999. Conformational difference between nuclear and cytoplasmic actin as detected by a monoclonal antibody. *J. Cell Sci.* **112**(Pt 6): 797–809.
26. Scheer, U. *et al.* 1984. Microinjection of actin-binding proteins and actin antibodies demonstrates involvement of

nuclear actin in transcription of lampbrush chromosomes. *Cell* **39:** 111–122.

27. Cairns, B.R. *et al.* 1998. Two actin-related proteins are shared functional components of the chromatin-remodeling complexes RSC and SWI/SNF. *Mol. Cell* **2:** 639–651.
28. Harata, M., R. Mochizuki & S. Mizuno. 1999. Two isoforms of a human actin-related protein show nuclear localization and mutually selective expression between brain and other tissues. *Biosci. Biotechnol. Biochem.* **63:** 917–923.
29. Harata, M. *et al.* 2000. Multiple actin-related proteins of Saccharomyces cerevisiae are present in the nucleus. *J Biochem.* **128:** 665–671.
30. Giesemann, T. *et al.* 1999. A role for polyproline motifs in the spinal muscular atrophy protein SMN. Profilins bind to and colocalize with smn in nuclear gems. *J. Biol. Chem.* **274:** 37908–37914.
31. Novak, K.D. & M.A. Titus. 1997. Myosin I overexpression impairs cell migration. *J. Cell Biol.* **136:** 633–647.
32. Pestic-Dragovich, L. *et al.* 2000. A myosin I isoform in the nucleus. *Science* **290:** 337–341.
33. Skare, P. *et al.* 2003. Profilin I colocalizes with speckles and Cajal bodies: a possible role in pre-mRNA splicing. *Exp. Cell Res.* **286:** 12–21.
34. Olave, I.A., S.L. Reck-Peterson & G.R. Crabtree. 2002. Nuclear actin and actin-related proteins in chromatin remodeling. *Annu. Rev. Biochem.* **71:** 755–781.
35. Pederson, T. & U. Aebi. 2005. Nuclear actin extends, with no contraction in sight. *Mol. Biol. Cell.* **16:** 5055–5060.
36. Rando, O.J., K. Zhao & G.R. Crabtree. 2000. Searching for a function for nuclear actin. *Trends Cell Biol.* **10:** 92–97.
37. Hegyi, G. *et al.* 1998. Intrastrand cross-linked actin between Gln-41 and Cys-374. I. Mapping of sites cross-linked in F-actin by N-(4-azido-2-nitrophenyl) putrescine. *Biochemistry (Mosc)* **37:** 17784–17792.
38. Hannappel, E., H. Kalbacher & W. Voelter. 1988. Thymosin β_4^{Xen}: a new thymosin β_4-like peptide in oocytes of Xenopus laevis. *Arch. Biochem. Biophys.* **260:** 546–551.

Ann. N.Y. Acad. Sci. ISSN 0077-8923

ANNALS OF THE NEW YORK ACADEMY OF SCIENCES
Issue: *Thymosins in Health and Disease*

Antibodies in research of thymosin β4: investigation of cross-reactivity and influence of fixatives

Jana Knop, Christine App, and Ewald Hannappel
Institute of Biochemistry, Friedrich-Alexander-University, Erlangen, Germany

Address for correspondence: Jana Knop, Institute of Biochemistry, Friedrich-Alexander-University, Fahrstr. 17, 91054 Erlangen, Germany. thymosin@biochem.uni-erlangen.de

Antibodies against thymosin β4 are available from various sources and have been used in immunohistochemistry, ELISA, and Western blot analyses. None of these antibodies have been fully characterized for specificity and influence of fixation techniques. This presents a difficulty because many tissues express more than one member of the β-thymosin family; in addition, highly homologous sequences are typical elements of β-thymosins. It is also important to scrutinize the influence of fixatives on the antibody-binding capability. Fixatives such as formaldehyde are well known as cross-linking reagents. Chemical modifications within the thymosin β4 molecule might change the putative epitope recognized by the antibody. These considerations suggest that investigations on thymosin β4 antibodies available to the scientific community are important and necessary before any experiment can be performed to exclude cross-reactivity with other β-thymosins that are coexistent in the examined tissue and to prove antibody binding after fixation steps.

Keywords: thymosin β4 antibody; cross-reactivity; cross-linking; fixative; formaldehyde

Introduction

Cross-reactivity is a reaction that can occur either between one antibody and various peptides and proteins or between an antigen and two or more antibodies. In research, these events are unintentional immunological reactions and, as a result, false positive reactions may be detected. Cross-reactivity of one antibody with several proteins or peptides is due to a similar amino acid sequence of the antigen. This similarity can either be caused by statistical probability or by proteins and peptides that belong to families with high amino acid homology. Thymosin β4 is known to be a member of the highly conserved β-thymosin family with very similar amino acid sequences.[1] This similar sequence can cause false positive results when antibodies against thymosin β4 are used. A linear epitope includes usually from five to eight amino acid residues on the surface of an antigen. The calculated statistical probability for the occurrence of an identical epitope consisting of five and eight amino acid residues is $1–3.2 \times 10^6$ and $1–2.5 \times 10^{10}$, respectively. It is not unlikely that in samples such as cell lysates containing many different proteins and peptides, antibody cross-reactions can occur. It is nearly impossible to exclude antibody cross-reactivity caused by statistical probability. However, cross-reactivity of an antibody with different targets belonging to the same highly homologous protein family are predictable and should be characterized to exclude false positive results.

Beside cross-reactivity, it is also important to examine the status of the epitope to exclude false negative results. For an efficient antigen–antibody interaction, the target epitope must be accessible for the antibody used. Modifications of the antigen might influence the target epitope and, therefore, abolish antibody binding. Epitope changes can occur after antigen denaturation triggered either by fixation, reduction, or the pH. These changes can cause the loss of conformation or even chemical modifications by fixatives. Nuclear magnetic resonance (NMR) data have shown that thymosin β4 is intrinsically unstructured and, therefore, pH changes

doi: 10.1111/j.1749-6632.2012.06659.x

should neither modify nor influence the antibody recognition site.[2] Treatment with reducing agents also do not influence the antibody binding because in β-thymosins no disulfide bridges are present. For chemical modifications of epitopes, the situation is different. For instance, because of the large number of lysine residues within the thymosin β4 molecule, fixatives with cross-linking properties can induce a variety of modifications within this peptide. Thus, putative changes in the epitope might interfere with the antibody recognition capability.

In this review, we summarize different methods to scrutinize the interaction of thymosin β4 antibodies with members of the β-thymosin family and His-tagged thymosin β4. In addition, different aspects of fixation steps in antibody binding are discussed.

Pretreatment of β-thymosins for antibody testing

Characterization of antibodies is possible using Western or dot blot experiments. The antibodies recognize epitopes of proteins or peptides adsorbed on a membrane. The adsorption depends on the molecular mass of the antigen, the strength of ionic and hydrophobic interactions with the surface of the membrane, water solubility, incubation conditions, and washing time.[3] Because of the low molecular mass of all β-thymosins (about 5 kDa), standard Western or dot blot analyses are not efficient for their detection. In these analyses, the small peptides are easily detached and dissolved from the membrane during the necessary washing steps. Increasing the molecular mass is one possibility to minimize this diffusion process and to enhance the adsorption to membranes.

Cross-linking reactions are useful tools to increase the molecular mass of β-thymosins. Homomolecular and heteromolecular cross-links can be formed. Homomolecular cross-links are formed between two or more identical molecules for instance after treatment of β-thymosins with formaldehyde or glutaraldehyde. Heteromolecular cross-links consist of different molecules and can be formed between β-thymosins and G-actin using 1-ethyl-3-(3-dimethylaminopropyl)carbodiimide (EDC).

Adsorption of β-thymosins on membranes increases after treatment with formaldehyde solutions due to various cross-linking reactions. In the case of thymosin β4, the N-terminus is acetylated. Neither cysteine nor tryptophan residues are present in its amino acid sequence. Therefore, only the ε-amino group of lysine residues is relevant for cross-linking reactions by formaldehyde. This cross-linking reaction consists of two steps. In a first step, formaldehyde reacts with the amino group to produce a carbinol that releases water and forms a Schiff base. In a second step, the Schiff base reacts with lysine side chains resulting in the formation of intramolecular or intermolecular methylen bridges (Fig. 1A).[4] Thymosin β4 is a tritetraconpeptide and contains nine lysine residues. This high number of lysine residues generates various possibilities for cross-linking reactions and might influence the interaction with antibodies.

Homomolecular cross-links of β-thymosin molecules can also be obtained by incubation of SDS gels (sodium dodecyl sulfate) in glutaraldehyde solution after electrophoresis.[5] Thus, Western blot experiments for the detection of thymosin β4 are possible. Intermolecular cross-links increase the molecular mass of thymosin β4 in the gel and therefore increase the adsorption to Western blot membranes during the transfer. Compared with formaldehyde, the cross-linking reaction of glutaraldehyde is of much more complex nature. Glutaraldehyde exists in aqueous solutions in multiple structures that might react with ε-amino groups of lysine residues. Because of the various polymeric structures of glutaraldehyde complex cyclic products can be generated through the cross-linking process.[6]

Heteromolecular cross-links can be performed using the ability of all β-thymosins to build a complex with G-actin.[7,8] Covalent bonds between G-actin and β-thymosins can be formed after treatment with cross-linking reagents. EDC is a zero-length cross-linker that leads to the formation of isopeptide bonds between the carboxyl group of aspartic or glutamic acid and the amino group of lysine residues.[9]

Influence of fixation techniques on recognition of thymosin β4 by antibodies

In immunohistochemistry, it is necessary to fixate the tissue to preserve morphological details. The most popular fixative is formaldehyde.[10] However, fixatives such as formaldehyde are cross-linking reagents that induce multiple chemical reactions within or between peptides.[11] This modifies the ε-amino group of lysine residues, which might

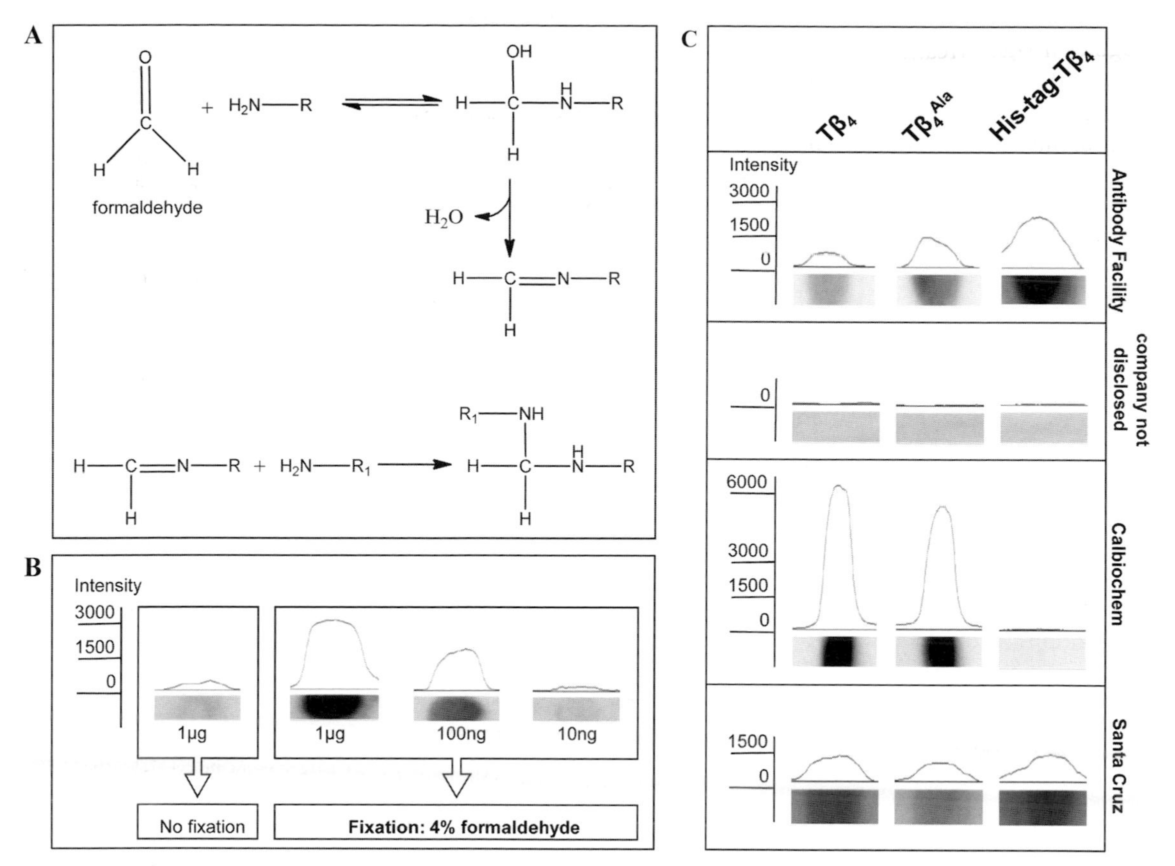

Figure 1. Fixation of β-thymosins using formaldehyde. (A) Reaction mechanism of fixation. Formaldehyde reacts with an amino group to produce a carbinol that releases water and forms a Schiff base. The Schiff base reacts with lysine side chains resulting in the formation of a methylen bridge. (B) Dot blot experiments of thymosin β4 applied on nitrocellulose membranes. Incubation of membranes with formaldehyde after applying thymosin β4 improves the sensitivity from 1 μg (no fixation) to 10 ng (after fixation) due to a better fixation. (C) Dot blot experiments to compare various thymosin β4 antibodies. One antibody did not recognize thymosin β4. All other antibodies showed cross-reactivity with thymosin β4^{Ala}. Only two of the antibodies were able to detect His-tagged thymosin β4.

significantly change the putative antigen epitope, especially in β-thymosins as every fifth amino acid residue is lysine. These modifications of the epitope might lead to a diminished or even abolished affinity of the antibody.

The influence of modifications of lysine residues in the antibody-binding site might be performed using dot blot experiments. When thymosin β4 is spotted on a nitrocellulose membrane, it is difficult to detect after the necessary blocking, washing, and antibody incubation steps. The peptide detaches from the membrane due to the low molecular mass of 5 kDa and the high water solubility of at least 100mg/mL. For fixation of β-thymosins, the spotted membrane was incubated in 3.5% buffered formaldehyde solution (according to 10% formalin solution, pH 7.3) as commonly used in immunohistochemistry.[12] Although fixation time usually takes 3–20 h, depending on the thickness of tissue sections, we fixated for only one hour at room temperature due to the direct exposure of the β-thymosins on the surface of the membrane. Formaldehyde treatment of membranes after applying thymosin β4 increases the peptide adsorption to the membrane. Owing to the fixation, a signal was obtained even with 10 ng of thymosin β4, while without fixation, an amount of 1 μg was necessary to obtain a comparable signal (Fig. 1B). These data demonstrate that treatment of thymosin β4 with formaldehyde does not interfere with the

	Human / Rat	Rabbit	Pig / Sheep	Cattle	Amino Acid Sequence (1 5 10 15 | 20 25 | 30 35 40)
$T\beta_4$	✓		✓	✓	ac-SDKPDMAEIEKFDKS KLKKTETQEKNPLP SKETIEQEKQAGES
$T\beta_4^{Ala}$		✓			ac-**A**DKPDMAEIEKFDKS KLKKTETQEKNPLP SKETIEQEKQAGES
$T\beta_9$				✓	ac-**A**DKPD**LG**EI**NS**FDK**A** KLKKTETQEKN**T**LP **T**KETIEQEKQA**K**
$T\beta_9^{Met}$			✓		ac-**A**DKPDM**G**EI**NS**FDK**A** KLKKTETQEKN**T**LP **T**KETIEQEK**RSEIS**
$T\beta_{10}$	✓	✓			ac-**A**DKPDM**G**EI**NS**FDK**A** KLKKTETQEKN**T**LP **T**KETIEQEK**RSEIS**
$T\beta_{15}$	✓				ac-SDKPD**LS**E**V**ETFDKS KLKKT**N**T**E**EKN**T**LP SKETI**Q**QEK**EYNQRS**
His-tag-$T\beta_4$					**H$_2$N**-SDKPDMAEIEKFDKS KLKKTETQEKNPLP SKETIEQEKQAGES**LEHHHHHH**
					actin-binding site (residues in the middle group)

Figure 2. Phylogenetic distribution and amino acid sequence of selected β-thymosins. Distribution of β-thymosins depends on species, tissue, and developmental stage. All β-thymosins contain highly homologues sequences; especially the actin-binding site is very similar. Amino acid residues differing from thymosin β4 are shown in bold letters.

recognition of the thymosin β4 antibodies tested. The fixation step before the blocking step in the dot blot experiment tremendously increases the sensitivity by preventing desorption of thymosin β4 from the membrane.

Cross-reactivity of thymosin β4 antibodies

Investigations on cross-reactivity of thymosin β4 antibodies with other β-thymosins are of utmost importance because of highly homologous amino acid sequences within the family (Fig. 2). The coexistence of β-thymosins in cells is dependent on species, tissue, and developmental stage. In humans, thymosin β4 and thymosin β10 are the main β-thymosins.[13,14] In addition, the expression of a third β-thymosin, thymosin β15, is increased in embryonic and carcinogenic tissue.[15] In rabbits, thymosin β4 is replaced by thymosin $\beta 4^{Ala}$ that differs only in the first amino acid residue where serine is replaced by alanine.[8] Thymosin β4 is also present in pigs, sheep, and calves, where it is accompanied by $\beta 9^{Met}$ in pigs and sheep or thymosin β9 in cattle.[16,17] The amino acid sequences of thymosin β9, $\beta 9^{Met}$, β10, and β15 differ only in a few amino acid residues at the N- and C-terminus when compared with thymosin β4 (Fig. 2). The central part of the peptides is highly conserved. This part is also considered as the G-actin binding site. Thymosin β4 antibodies should be able to distinguish between the β-thymosins that are coexisting in the examined tissue. Unfortunately, there are almost no data available about cross-reactivity with other members of the β-thymosin family.

To study the cross-reactivity, it is necessary to isolate β-thymosins from spleen of different species.[18] It is advisable to cross-link the isolated β-thymosins to increase the adsorption to membranes. This can be achieved by heteromolecular cross-linking of β-thymosins to G-actin by EDC (Fig. 3A). The cross-linked product, formed between one molecule of β-thymosin and one molecule G-actin, can be separated using SDS gel electrophoresis (Fig. 3B). Cross-reactivity of thymosin β4 antibodies has been tested by Western blot analysis after SDS gel electrophoresis of different β-thymosins cross-linked to G-actin.

Our review on cross-reactivity includes four different commercially available antibodies raised in different hosts and against different epitopes (Table 1). One antibody was of monoclonal nature, the other three were polyclonal. It turned out that one of the investigated antibodies (company not disclosed) was not working in this or any other experiment. This fact demonstrates the importance of antibody testing before any experiment can be performed. The other three antibodies recognized the cross-linked thymosin β4 in the Western blot as expected, though they also showed a signal for thymosin $\beta 4^{Ala}$. None of these antibodies showed cross-reactivity with thymosin β9, thymosin $\beta 9^{Met}$, and thymosin β10 (Fig. 3C).

Cross-reactivity of thymosin β4 antibodies with thymosin $\beta 4^{Ala}$ might be relevant in experiments

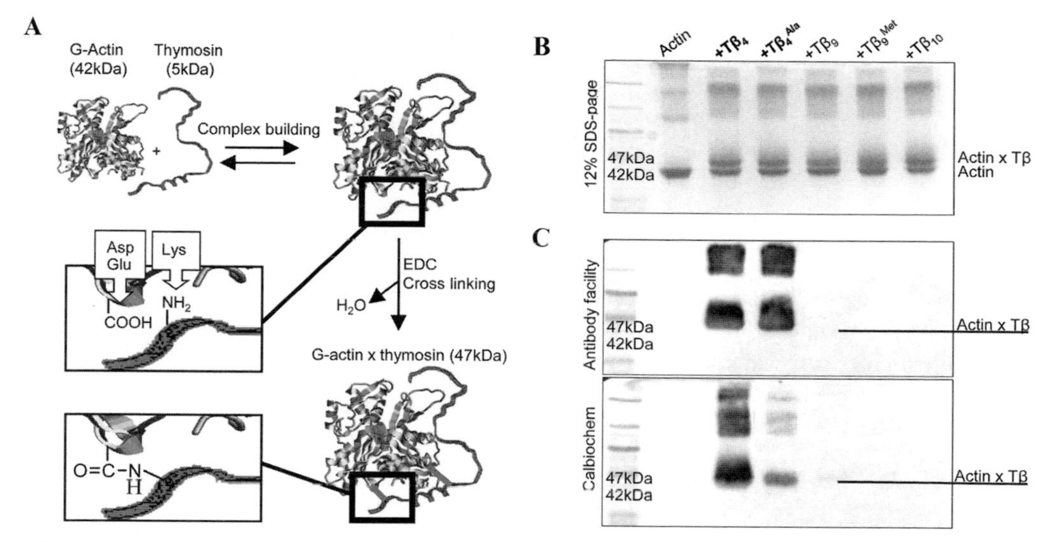

Figure 3. Cross-linking of β-thymosins to G-actin by EDC. (A) Schematic presentation of cross-linking reaction of thymosin β4 to G-actin. (B) SDS-polyacrylamide gel electrophoresis after EDC treatment of G-actin in the absence (lane 2) or presence of various β-thymosins (lanes 3–7); molecular weight standard (lane 1). Non-cross-linked G-actin is represented by a band at 42 kDa. Cross-linked products with β-thymosins display a second band at 47 kDa. Bands at higher molecular masses are caused by cross-links between two or more G-actin molecules. (C) Western blot analyses of cross-linked products with two different thymosin β4 antibodies. Both antibodies show cross-reactivity with thymosin $\beta4^{Ala}$.

using rabbit cells that have been transfected with DNA encoding for thymosin β4. In this case, the probability of false positive detection of expressed thymosin β4 caused by cross-reactivity with naturally occurring thymosin $\beta4^{Ala}$ must be excluded. According to our knowledge, no antibodies are available that can distinguish between thymosin β4 and thymosin $\beta4^{Ala}$. Currently, it is only possible to distinguish these two peptides either by RP–HPLC or mass spectrometry.[19]

It is also possible to use dot blot experiments to analyze cross-reactivity and specificity of thymosin β4 antibodies. In our group, the specificity of thymosin β4 antibodies to thymosin β4 and thymosin $\beta4^{Ala}$ was analyzed using formaldehyde fixation on membranes (Fig. 1C). Thymosin β4 antibodies were able to bind to thymosin β4 as expected but also showed cross-reactivity with thymosin $\beta4^{Ala}$. These data are consistent with the investigation on cross-reactivity using Western blot analysis with β-thymosins cross-linked to G-actin.

Although the dot blot procedure seems to be the easiest method for antibody testing, it is necessary to consider that multiple unknown cross-linking reactions by formaldehyde might change the putative epitope. It is possible that formaldehyde leads to the modification of all nine lysine residues within the thymosin β4 molecule. This could be of importance if the epitope contains one or more of these lysine residues. Another option to increase the adsorption

Table 1. Information on thymosin β4 antibodies given by companies[a]

Antibody	Company	Immunogen	Host	Clonality	Reported cross-reactivity
Anti-thymosin β4 5H11	Antibody facility	Tβ4 from horse	mouse	monoclonal	no data available
Thymosin β4-antiserum	not disclosed	Tβ4	rabbit	polyclonal	no data available
Anti-thymosin β4 rabbit	Calbiochem	Tβ4(1–14)	rabbit	polyclonal	no cross-reactivity with thymosin α1 or thymulin
Tβ4(N-18) antibody	Santa Cruz	Tβ4(1–18)	goat	polyclonal	no data available

[a]No data about cross-reactivity with other β-thymosins are available.

to membranes is cross-linking β-thymosins to G-actin by EDC. This reaction forms only two isopeptide bonds between β-thymosin and G-actin and might not influence the antibody affinity (Fig. 3A).[9]

Specificity of thymosin β4 antibodies to His-tagged thymosin β4

In contrast to the provided high specificity, the thymosin β4 antibodies should be able to detect tagged forms of thymosin β4 expressed in transfected cells. In the case of His-tagged thymosin β4, the missing acetylation of the N-terminus and additional amino acids at the C-terminus may change epitopes recognized by antibodies (Fig. 2). The proof of antibody specificity to tagged forms of thymosin β4 is from the same problematic nature as nontagged forms. Small molecular mass complicates classical Western or dot blot experiments. This tagged form of thymosin β4 is also able to form a complex with G-actin and can be cross-linked to G-actin by EDC (data not shown). Antibody specificity might be evaluated after gel electrophoresis and Western blot analyses. Another option to investigate the antibody specificity is the fixation of tagged thymosin β4 on membranes with formaldehyde. Only two of the commercially available antibodies were able to bind to the His-tagged form of thymosin β4 (Fig. 1C), while the antibody from Calbiochem (Darmstadt, Germany) was able to distinguish between thymosin β4 and His-tagged thymosin β4.

Conclusion

In summary, the treatment of thymosin β4 with formaldehyde does not interfere with its recognition by specific antibodies. The treatment leads to a more sensitive detection of thymosin β4 due to a better fixation on the membrane. SDS gel electrophoresis of β-thymosins cross-linked to G-actin and Western blot analysis as well as dot blot analysis of β-thymosins fixated by formaldehyde are easy methods to evaluate the specificity of antibodies. This test is necessary before any experiment can be performed to exclude cross-reactivity with other members of the β-thymosin family that may coexist in the examined tissue. Thymosin β4 antibodies showed cross-reactivity with thymosin $\beta4^{Ala}$ but did not bind to thymosin β9, thymosin $\beta9^{Met}$, and thymosin β10. C-terminally His-tagged thymosin β4 was recognized only by two of the commercially available thymosin β4 antibodies.

Acknowledgment

The authors want to thank Prof. H. G. Mannherz and D. Jaegers for support, as well as J. Arellanes-Robledo for comments on the manuscript.

Conflicts of interest

The authors declare no conflicts of interest.

References

1. Huff, T., C.S. Muller, A.M. Otto, *et al.* 2001. Beta-thymosins, small acidic peptides with multiple functions. *Int. J. Biochem. Cell Biol.* **33:** 205–220.
2. Zarbock, J., H. Oschkinat, E. Hannappel, *et al.* 1990. Solution conformation of thymosin beta 4: a nuclear magnetic resonance and simulated annealing study. *Biochemistry (Mosc).* **29:** 7814–7821.
3. Pristoupil, T.I., M. Kramlova & J. Sterbikova. 1969. On the mechanism of adsorption of proteins to nitrocellulose in membrane chromatography. *J. Chromatogr.* **42:** 367–375.
4. Toews, J., J.C. Rogalski, T.J. Clark, *et al.* 2008. Mass spectrometric identification of formaldehyde-induced peptide modifications under in vivo protein cross-linking conditions. *Anal. Chim. Acta* **618:** 168–183.
5. Ho, J.H., C.H. Chuang, C.Y. Ho, *et al.* 2007. Internalization is essential for the antiapoptotic effects of exogenous thymosin beta-4 on human corneal epithelial cells. *Invest. Ophthalmol. Vis. Sci.* **48:** 27–33.
6. Migneault, I., C. Dartiguenave, M.J. Bertrand, *et al.* 2004. Glutaraldehyde: behavior in aqueous solution, reaction with proteins, and application to enzyme crosslinking. *Biotechniques* **37:** 790–796, 798–802.
7. Safer, D., M. Elzinga & V.T. Nachmias. 1991. Thymosin beta 4 and fx, an actin-sequestering peptide, are indistinguishable. *J. Biol. Chem.* **266:** 4029–4032.
8. Huff, T., C.S. Muller & E. Hannappel. 2007. Thymosin beta4 is not always the main beta-thymosin in mammalian platelets. *Ann. N. Y. Acad. Sci.* **1112:** 451–457.
9. Safer, D., T.R. Sosnick & M. Elzinga. 1997. Thymosin beta 4 binds actin in an extended conformation and contacts both the barbed and pointed ends. *Biochemistry (Mosc).* **36:** 5806–5816.
10. Werner, M., A. Chott, A. Fabiano, *et al.* 2000. Effect of formalin tissue fixation and processing on immunohistochemistry. *Am. J. Surg. Pathol.* **24:** 1016–1019.
11. Metz, B., G.F. Kersten, P. Hoogerhout, *et al.* 2004. Identification of formaldehyde-induced modifications in proteins: reactions with model peptides. *J. Biol. Chem.* **279:** 6235–6243.
12. Namimatsu, S., M. Ghazizadeh & Y. Sugisaki. 2005. Reversing the effects of formalin fixation with citraconic anhydride and heat: a universal antigen retrieval method. *J. Histochem. Cytochem.* **53:** 3–11.

13. Erickson-Viitanen, S., S. Ruggieri, P. Natalini, *et al.* 1983. Thymosin beta 10, a new analog of thymosin beta 4 in mammalian tissues. *Arch. Biochem. Biophys.* **225:** 407–413.
14. Hannappel, E. 2010. Thymosin beta4 and its posttranslational modifications. *Ann. N. Y. Acad. Sci.* **1194:** 27–35.
15. Banyard, J., C. Barrows & B.R. Zetter. 2009. Differential regulation of human thymosin beta 15 isoforms by transforming growth factor beta 1. *Genes. Chromosomes Cancer* **48:** 502–509.
16. Hannappel, E., S. Davoust & B.L. Horecker. 1982. Thymosins beta 8 and beta 9: two new peptides isolated from calf thymus homologous to thymosin beta 4. *Proc. Natl. Acad. Sci. USA* **79:** 1708–1711.
17. Hannappel, E., F. Wartenberg & X.R. Bustelo. 1989. Isolation and characterization of thymosin beta 9 met from pork spleen. *Arch. Biochem. Biophys.* **273:** 396–402.
18. Hannappel, E., S. Davoust & B.L. Horecker. 1982. Isolation of peptides from calf thymus. *Biochem. Biophys. Res. Commun.* **104:** 266–271.
19. Huff, T., C.S.G. Müller & E. Hannappel. 1997. Hplc and postcolumn derivatization with fluorescamine. Isolation of actin-sequestering β-thymosins by reversed-phase hplc. *Anal. Chim. Acta* **352:** 239–248.

Ann. N.Y. Acad. Sci. ISSN 0077-8923

ANNALS OF THE NEW YORK ACADEMY OF SCIENCES
Issue: *Thymosins in Health and Disease*

Thymosin β4 sustained release from poly(lactide-co-glycolide) microspheres: synthesis and implications for treatment of myocardial ischemia

Jeffrey E. Thatcher,[1,2] Tré Welch,[3] Robert C. Eberhart,[2,3] Zoltan A. Schelly,[4] and J. Michael DiMaio[1]

[1]Department of Cardiothoracic Surgery, University of Texas Southwestern Medical Center, Dallas, Texas. [2]Department of Biomedical Engineering, University of Texas, Arlington, Texas. [3]Department of Surgery, University of Texas Southwestern Medical Center, Dallas, Texas. [4]Department of Chemistry and Biochemistry, University of Texas, Arlington, Texas

Address for correspondence: Jeffrey Thatcher, Department of Cardiovascular and Thoracic Surgery, The University of Texas Southwestern Medical Center, 5323 Harry Hines Blvd., Dallas, TX 75390, Mail Code: 8879. Jeffrey.Thatcher@UTSouthwestern.edu

A sustained release formulation for the therapeutic peptide thymosin β4 (Tβ4) that can be localized to the heart and reduce the concentration and frequency of dose is being explored as a means to improve its delivery in humans. This review contains concepts involved in the delivery of peptides to the heart and the synthesis of polymer microspheres for the sustained release of peptides, including Tβ4. Initial results of poly(lactic-co-glycolic acid) microspheres synthesized with specific tolerances for intramyocardial injection that demonstrate the encapsulation and release of Tβ4 from double-emulsion microspheres are also presented.

Keywords: controlled; sustained; release; microsphere; thymosin β4; heart

Introduction

The need for advancements in delivery of biological molecules that treat heart disease, such as the peptide thymosin β4 (Tβ4), is increasing rapidly. Many attempts in humans and large animals reveal that peptides do not have significant effects when delivered intravascularly, owing to short serum half-life, poor tissue uptake, and systemic side effects.[1–5] More successful attempts to deliver peptides to the heart include direct myocardial injection and injection into the pericardial space.[6,7] Although these types of procedures are feasible, they are invasive and less amenable to multiple-dosing regimens.[8] Previous studies such as these are motivating the development of a sustained or controlled-release system customized to the molecular composition and biological effects of Tβ4. This review will focus on biodegradable microspheres as a means for sustained release of Tβ4 in the heart, including some early results of Tβ4 release with this technique.

Tβ4 as a therapy for ischemic heart disease

The cellular mechanisms influenced by Tβ4 establish the location and duration of administration to the heart during myocardial ischemia. These mechanisms occur in cardiomyocytes and epithelial cells of the epicardium. The result of their activation by Tβ4 is the lengthening of the duration of cardiomyocyte survival and the induction of the formation of new vasculature by mobilizing cells from the epicardial layer.[9,10]

The therapeutic benefits of Tβ4 occur when the peptide is administered over a duration that spans the major postischemia cellular events in the myocardium. This timeframe includes cardiomyocyte apoptosis in the ischemic region as well as fibroblast infiltration and collagen synthesis. Although it has not been explored, there may be a therapeutic benefit of administration of Tβ4 throughout cardiac remodeling where apoptosis and collagen synthesis occur for weeks to months in the region of

doi: 10.1111/j.1749-6632.2012.06681.x

myocardium surrounding the initial infarct, the "border zone."[11,12]

The dosing regimen used to achieve a therapeutic response in the myocardium in adult mice (25–30 g) is 6 mg/kg intraperitoneal (IP) at the time of injury and every three days for up to four weeks.[9,13] In addition, delivery of 400 ng Tβ4 intramyocardially at time of injury with or without parallel IP dosing also results in significant functional improvement.[9] In another model of ischemia, a rat stroke model, 6 mg/kg Tβ4 given IP every three days for four additional doses is effective for reducing cognitive decline by what is hypothesized to be axonal remodeling at the ischemic boundary.[14] The cardiac dosing regimens are supported by a pharmacokinetic study that demonstrates that the heart has a significant increase in Tβ4 concentration compared to baseline (80 μg/g vs. 37 μg/g) when Tβ4 is delivered IP at approximately 16 mg/kg.[15] In this study, peak increases in Tβ4 concentration occurred at two hours and diminished almost to baseline 24 h after administration.

Routes of delivery in humans include intravascular, IP, intramyocradial, and injection into the pericardial space. In direct myocardial injection, prevention of injectate efflux is one of the largest hurdles,[16] which may be overcome by reducing injectate volume by delivering the therapeutic within a viscous material or an implantable matrix or by slowing the rate of injection. Formulations for sustained release may also be delivered into the pericardial space via the intrapericardial route. By this route, peptide diffusion from the pericardial space to the subendocardium is minimal.[17,18] However, this method may have a potent effect on the epicardium and could have excellent long-term results in combination with direct myocardial injection. The invasiveness of these methods of delivery lend themselves to the use of a single-dose formulation that remains at the site of injection and slowly releases drug for an extended duration.

Controlled-release technologies for drug delivery to the heart

Controlled-release matrices that contain Tβ4 demonstrated retention of therapeutic activity over extended durations and induced a robust angiogenic response in endothelial cells. Two such applications have shown that Tβ4 can induce vascular gene expression and organization in endothelial cells within a network of hydrophilic polymer, a hydrogel. In the first case, Tβ4 was immobilized in a poly(ethylene glycol) (PEG) matrix that contained matrix metalloproteinase (MMP) cleavable cross-linking peptides and integrin adhesion peptides for cell attachment.[19] In the second example, a collagen matrix combined with chitosan was assembled to function as a controlled-release matrix for Tβ4, as well as a tissue-filling agent to reduce the amount of cardiac remodeling after myocardial infarction.[20] These studies in Tβ4 sustained release reiterate the effect of Tβ4 as a potent angiogenic factor. One of the major improvements in Tβ4 delivery by controlled-release technology was the retention of therapeutically active peptide over an extended duration, which resulted in a more potent biological response.

The effectiveness of controlled-release formulations has been successfully demonstrated in the heart using animal models. Myocardial injections of growth factors and small molecules have been performed in animal models using various controlled-release matrices. In general, these show improved outcomes compared to bolus injection. The materials used in these studies include gelatin hydrogel microspheres,[21–24] self-assembling peptide nanofibers,[25] and alginate hydrogels.[26] Hsieh[25] and Hao[26] demonstrated their controlled-release matrices lead to physiologically active drug levels for at least 14 and 15 days after injection, respectively. These are much improved therapeutic retentions compared to bolus injection. Clinical trials have assessed the administration of a heparin-alginate slow-release matrix implanted perivascularily in the epicardial fat in humans.[27] Results from this small study demonstrate that injectable, controlled-release delivery systems are safe in patients and may be effective therapy to treat ischemic myocardium that cannot be reperfused by a graft.

Controlled release of peptides from polymer matrices

Controlled release of macromolecules such as proteins and peptides has been achieved through entrapment in a variety of polymers including: poly(amides), poly(amino acids), poly(alkyl-a-cyanoacrylates), poly(esters), poly(orthoesters), poly(urethanes), and poly(acrylamides).[28] This technology has been applied to a multitude of pharmaceutical applications, such as the release of antigens for immunization,[29,30] peptides to promote tissue regeneration,[31,32] and hormone replacement therapy.[33,34]

We will narrow our focus to poly(lactic-co-glycolic acid) (PLGA), a biodegradable polyester composed of lactic acid (LA) and glycolic acid (GA) monomers. The advantage in using LA and GA derived polymers for drug delivery comes from their biocompatibility and customizability. For PLGA controlled-release devices, the tissue is generally compatible with this polymer and its acidic degradation products.[35,36] Furthermore, these polymers have the advantage of being stable, accessible, and the most widely studied biodegradable polymer. For this reason, an accurate estimation of the required molecular weight (M_w) and monomer ratio is possible.

Drug release from these polymers occurs by a complex mechanism that includes diffusion, polymer degradation, and drug-polymer interactions. Finally, the activity of the released drug is affected by the processing parameters in device synthesis as well as changes that occur to the matrix after being rehydrated in the tissue. Drug interactions with the polymer and its degradation products, other drug molecules, and additional excipients in the formulation add to the complexity of the system. For these reasons, formulations are numerous and customized to specific therapeutic molecules to achieve desirable encapsulation and release properties.

One particularity useful and well-studied configuration for controlled release of biomolecules is the microsphere. Microspheres are small polymer beads with sizes ranging from 1 to 500 μm in diameter. Their size and shape makes them amenable to delivery by injection, and formulations can be tailored to release one or multiple therapeutics over durations from days to months.

LA-derived polyesters for peptide release

Polyesters of LA and/or GA are biodegradable by the scission of the ester linkages in the presence of water. The reaction can be catalyzed in the presence of acids, bases, salts, or enzymes.[37] Common polyesters for microsphere synthesis include poly(lactic acid) (PLA), poly(glycolic acid) (PGA), PLGA, PLGA-poly(ethylene glycol) block copolymers.[38–41] PLA, PGA, and their copolymers are hydrophobic. PLA is semi-crystalline when synthesized from one stereoisomer configuration of LA (either the *d*- or the *l*- form). These crystalline regions make the polymer more resistant to water penetration resulting in degradation over approximately two years at physiological pH. Polymers made from both *d*- and *l*- LA isoforms are amorphous, because the racemic polymer does not arrange into crystalline regions. Lack of crystallinity gives water easier access to hydrolysable regions of the polymer chain shortening degradation time to 12–16 months. PGA is a highly crystalline polymer that is more hydrophilic than PLA. Owing to its hydrophilicity, it does not maintain mechanical strength compared to PLA and degrades more rapidly, in two to three months. These degradation rates are influence by polymer M_w.[42]

Copolymerization of LA with GA results in a polymer with many beneficial characteristics for drug delivery. This polymer is amorphous, making it a good substrate for drug entrapment. PLGA microspheres degrade much more quickly than PLA microspheres.[43] One of the advantages of this system is that the microsphere degradation rate can be modified from approximately one to six months by altering the LA:GA ratio.[42] Microspheres fabricated from PLGA 50:50 degrade slightly faster than those from PLGA 75:25, both of which degrade much faster than PLA.[44]

Peptide release can also be altered by manipulating the LA:GA ratio. For instance, PLGA 75:25 can release nerve growth factor more quickly than PLGA 50:50,[45] bovine serum albumin was released faster in PLGA 50:50 than in PLGA 75:25 or 85:15,[46] and finally melittin was released more quickly in PLGA 50:50 than PLGA 75:25.[44] These results support the effect of modulating the copolymer ratio to alter drug release rate. However, the most rapidly degradable polymer is not always the fastest peptide-releasing polymer, likely a result of drug-polymer interactions or the heterogeneous process of microsphere degradation described later.

Polymer M_w affects the mass-loss of polyester microspheres. Microsphere mass-loss results from polymer chain scission and the subsequent diffusion of oligomers and monomers from the bulk of the sphere. Microspheres fabricated by double emulsion from PLGA of M_w 12,000 Daltons (Da) will completely disintegrate in one month (from spheres to polymer mass), whereas microspheres of M_w 30,000

Da retain their integrity over the same duration.[47] In general, drug release rates from polyesters are reduced for polymers of increasing M_w.[44] However, microspheres synthesized from higher M_w PLGA (40,000–80,000 Da) can release protein in short durations based on the method of preparation.[48–51] Therefore, it is more important to choose polymer M_w for polymer residence time and rate of monomer formation than for its effect on drug release.

Microsphere synthesis

There are various techniques used to synthesize peptide-containing microspheres. To avoid damaging temperature sensitive drugs, the polymer is often dissolved in an organic solvent at or near room temperature instead of being melted to for the desired shape. This polymer-containing solvent is then dispersed in an immiscible phase (e.g., aqueous or gas) to create spherical droplets followed by evaporation of the solvent. Two common methods are water-in-oil-in-water double emulsion, and solid-in-oil-in-water emulsion. These methods can be applied to work with different polymer-peptide combinations. A more detailed review of microsphere synthesis techniques has been provided.[28]

We prepared microspheres using the double-emulsion technique from PLGA (M_w 41–59 kDa; Fig. 1). The formulation contained the surfactants octyl-glucopyranosid or decyl-glucopyranoside in the organic phase (5% PLGA [w:v] in dichloromethane) at a concentration of 8.55 mM. These biocompatible surfactants were incorporated into our formulation to influence the rate of peptide release. The aqueous phase contained 2.5% Tβ4 (w:v). Encapsulation of Tβ4 was very efficient in our formulations: 74 ± 14 (% ± SD) in unblended microspheres, 98 ± 1 in PLGA blended with octyl-glycopyranoside, and 82 ± 6 in PLGA blended with decyl-glucopyranoside. The overall high encapsulation efficiency may be the result of the frequency of positively charged residues on Tβ4. These groups may have prevented peptide loss during microsphere synthesis by promoting electrostatic interactions with the negatively charged PLGA.

Retention of the injectate for cardiac applications requires that the microspheres be larger than the capillaries of the myocardium so they will not be

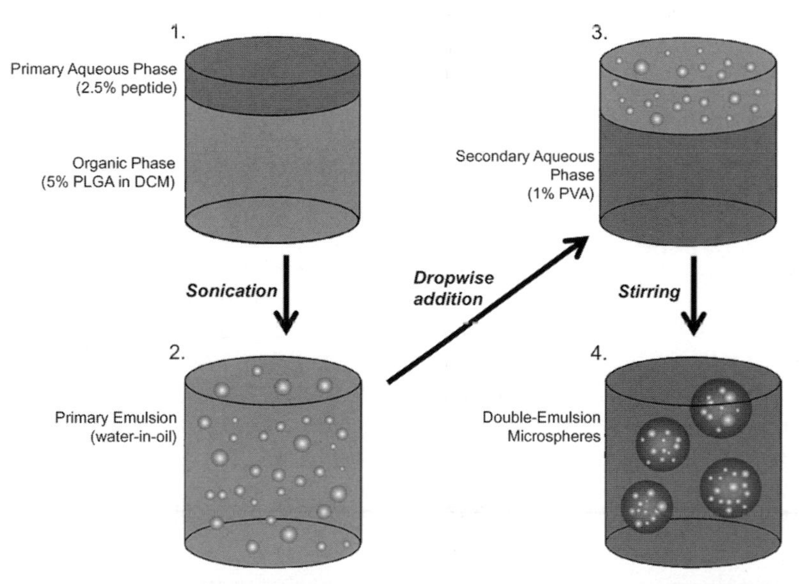

Figure 1. The double-emulsion method demonstrating synthesis of Tβ4-containing microspheres. (1) In step one, the aqueous phase containing Tβ4 was added to the organic phase, dichloromethane (DCM), containing PLGA. The ratio of organic phase to aqueous phase was 1:25 (v:v). (2) In the second step, the two-phase mixture generated in step one was mixed with high-intensity sonication to create an emulsion with water droplets less than 1 μm in diameter. (3) Step three was the addition of the primary emulsion to an aqueous solution containing a small amount of stabilizing surfactant, poly(vinyl alcohol) (PVA). The ratio of the primary emulsion to the secondary aqueous phase was 1:4 (v:v). By gentle stirring with an impeller or magnetic stirring bar at 300 rpm, double-emulsion microspheres appeared. (4) Finally, the organic solvent was evaporated from the microspheres as stirring continued.

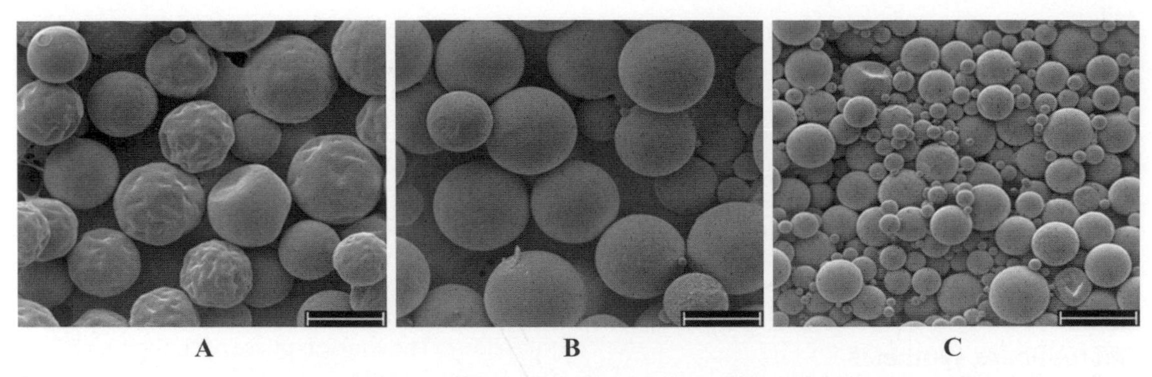

Figure 2. Scanning electron microscopy images of PLGA microspheres containing Tβ4. Scale bar represents 200 μm. Microspheres were prepared with or without the addition of a surfactant blended with the polymer in the organic phase. (A) Microspheres without surfactant; (B) microspheres prepared with octyl-glycopyranoside blended with PLGA; (C) microspheres with decyl-glucopyranoside blended with PLGA. The average diameters of the formulations were (A) 182 ± 63 (μm ± SD); (B) 178 ± 79; (C) 72 ± 44. Microspheres blended with decyl-glucopyranoside were significantly smaller ($P < 0.001$; ANOVA).

easily swept away by local circulation. Intravascular perfusion of radiolabeled microspheres in the excised dog or cat heart indicate their smallest capillaries are between 8.6 and 14.6 μm.[52] To ensure our PLGA microspheres did not contain particles below this specific size range we washed the final preparation over a 15 μm sieve. After this step, we observed microspheres in our formulation to be greater than 20 μm in all three of our Tβ4 formulations (Fig. 2). Although this method may improve microsphere retention, it is likely that a small portion of the spheres will flow through veins in the myocardium and become lodged in the lungs. This complication may be avoided by delivering the microspheres within a viscous vehicle such as a hydrogel.

Degradation and peptide release from PLGA microspheres

Microspheres composed of PLGA degrade by a process called bulk degradation, where the rate of water penetration far exceeds the rate of PLGA degradation.[53] Therefore, the entire microsphere is wetted before any significant polymer scission. Hydrolysis occurring within the microsphere leads to formation of acidic PLGA oligomers. The retention of acidic oligomers influences the degradation mechanism of the microspheres.

Insoluble PLGA oligomers perpetuate the degradation of a PLGA microsphere, which is a heterogeneous process occurring faster toward the center. Faster polymer degradation in the inside of the microsphere is initially attributable to the inability of polymer strands in the interior of the microsphere to diffuse outward. This causes an elevated concentration of acidic degradation products that increase the rate of PLGA hydrolysis. This increased rate in the presence of degradation products is called autocatalysis.[54] The morphological result of autocatalysis is the appearance of a hollow microsphere interior.

Occurring simultaneously with the hollowing of the interior is the formation of a thick shell of crystalline polymer on the microsphere surface. The thick shell also influences autocatalysis by trapping small and acidic water-soluble PLGA degradation products in the interior. This interior environment of low pH[55] and high concentration of polymer degradation products may cause covalent attachment of GA or LA monomers to labile amino acids on the entrapped peptide.[56] These effects are troublesome for peptide delivery, yet may be mitigated by addition of protective molecules.[57]

The heterogeneous degradation process is typically, what leads to a commonly reported three-phase release of drug. Initially, a drug at or near the surface of the newly wetted microspheres can quickly escape. This initial release occurs within the first 24 h and is called *burst release*. Between 24 h and approximately two weeks, the thickening of the outer shell takes place, which inhibits drug encapsulated in the interior of the microsphere from escaping. This phase of very little drug release is called the lag phase. Eventually the outer shell thins and becomes porous, allowing drug trapped within to escape. This is the continued release phase.

Our microsphere formulations demonstrated these three phases when Tβ4 was released from

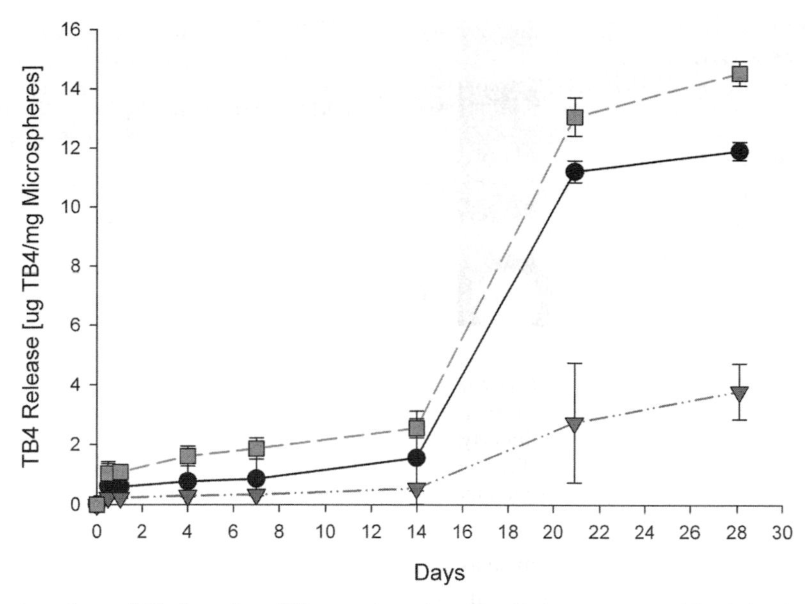

Figure 3. The *in vitro* release of Tβ4 from three different microsphere formulations. PLGA was blended with surfactants in these formulations: o, unblended polymer; □, PLGA blended with octyl-glucopyranoside; Δ, PLGA blended with decyl-glucopyranoside. Decyl-glucopyranoside–blended microspheres release the drug significantly slower than both unblended and octyl-glucopyranoside blended formulations ($P < 0.017$; RM-ANOVA). Percentages to the right show the total amount of Tβ4 released from the total amount of Tβ4 loaded into the microsphere formulation after 28 days.

microspheres using phosphate buffered saline (pH 7.4) at 37 °C as the release medium (Fig. 3). We found a short initial burst-release phase followed by a lag phase lasting for two weeks. The final phase of drug release beginning at 14 days resulted in the highest release of drug. This information is vital to our understanding of how certain factors in microsphere preparation contribute to the release of Tβ4. However, *in vitro* drug release experiments poorly mimic the environment in the myocardium during ischemia. In the pathological state, this tissue contains many factors that may enhance the rate of microsphere degradation, such a decreased pH[58] and the presence of oxygen radicals,[59] which may significantly shorten the 14-day lag period observed *in vitro*.

Summary

Sustained release formulation for delivery of Tβ4 that can be easily administered to the myocardium will reduce the number of invasive interventions, protect the drug from degradation, and localize the dose to the target cardiomyocytes and epicardial cells. Formulations for sustained release of Tβ4 applied to different anatomical regions around the heart may provide more advantageous drug dosing. Results with Tβ4-containing polymer microspheres are promising in that they provide high encapsulation of peptide, and can be fashioned within a range of sizes amenable for parenteral routes of delivery. The lag phase of Tβ4 release from microspheres occurs during especially important pathophysiologic events where exposure of the myocardium to Tβ4 is necessary. *In vivo* drug release will be necessary to determine whether formulation characteristics of our microspheres will need to be adjusted to release Tβ4 more rapidly or be combined with a vehicle capable of delivering a rapid initial dose of Tβ4.

Acknowledgment

The authors thank Ildiko Bock-Marquette for her scientific guidance and friendship.

Conflicts of interest

The authors declare no conflicts of interest.

References

1. Aiello, L.P., R.L. Avery, P.G. Arrigg, *et al.* 1994. Vascular endothelial growth factor in ocular fluid of patients with diabetic retinopathy and other retinal disorders. *N. Engl. J. Med.* **331:** 1480–1487.
2. Cooper, L.T., Jr., W.R. Hiatt, M.A. Creager, *et al.* 2001. Proteinuria in a placebo-controlled study of basic fibroblast

growth factor for intermittent claudication. *Vasc. Med.* **6:** 235–239.
3. Jain, R.K., A.V. Finn, F.D. Kolodgie, *et al.* 2007. Antiangiogenic therapy for normalization of atherosclerotic plaque vasculature: a potential strategy for plaque stabilization. *Nat. Clin. Pract. Cardiovasc. Med.* **4:** 491–502.
4. Lucerna, M., A. Zernecke, R. de Nooijer, *et al.* 2007. Vascular endothelial growth factor-A induces plaque expansion in ApoE knock-out mice by promoting de novo leukocyte recruitment. *Blood* **109:** 122–129.
5. Virmani, R., F.D. Kolodgie, A.P. Burke, *et al.* 2005. Atherosclerotic plaque progression and vulnerability to rupture: angiogenesis as a source of intraplaque hemorrhage. *Arterioscler. Thromb. Vasc. Biol.* **25:** 2054–2061.
6. Laham, R.J., M. Post, M. Rezaee, *et al.* 2005. Transendocardial and transepicardial intramyocardial fibroblast growth factor-2 administration: myocardial and tissue distribution. *Drug Metab. Dispos.* **33:** 1101–1107.
7. Lazarous, D.F., M. Shou, J.A. Stiber, *et al.* 1997. Pharmacodynamics of basic fibroblast growth factor: route of administration determines myocardial and systemic distribution. *Cardiovasc Res.* **36:** 78–85.
8. Fuchs, S., L.F. Satler, R. Kornowski, *et al.* 2003. Catheter-based autologous bone marrow myocardial injection in no-option patients with advanced coronary artery disease: a feasibility study. *J. Am. Coll. Cardiol.* **41:** 1721–1724.
9. Bock-Marquette, I., A. Saxena, M.D. White, *et al.* 2004. Thymosin beta4 activates integrin-linked kinase and promotes cardiac cell migration, survival and cardiac repair. *Nature* **432:** 466–472.
10. Bock-Marquette, I., S. Shrivastava, G.C. Pipes, *et al.* 2009. Thymosin beta4 mediated PKC activation is essential to initiate the embryonic coronary developmental program and epicardial progenitor cell activation in adult mice in vivo. *J. Mol. Cell Cardiol.* **46:** 728–738.
11. Mani, K. & R.N. Kitsis. 2003. Myocyte apoptosis: programming ventricular remodeling. *J. Am. Coll. Cardiol.* **41:** 761–764.
12. Sun, Y., J.Q. Zhang, J. Zhang & S. Lamparter. 2000. Cardiac remodeling by fibrous tissue after infarction in rats. *J. Lab. Clin. Med.* **135:** 316–323.
13. Smart, N., C.A. Risebro, J.E. Clark, *et al.* 2010. Thymosin beta4 facilitates epicardial neovascularization of the injured adult heart. *Ann. N. Y. Acad. Sci.* **1194:** 97–104.
14. Morris, D.C., M. Chopp, L. Zhang, *et al.* 2010. Thymosin beta4 improves functional neurological outcome in a rat model of embolic stroke. *Neuroscience* **169:** 674–682.
15. Mora, C.A., C.A. Baumann, J.E. Paino, *et al.* 1997. Biodistribution of synthetic thymosin beta 4 in the serum, urine, and major organs of mice. *Int. J. Immunopharmacol.* **19:** 1–8.
16. Grossman, P.M., Z. Han, M. Palasis, *et al.* 2002. Incomplete retention after direct myocardial injection. *Catheter Cardiovasc. Interv.* **55:** 392–397.
17. Gleason, J.D., K.P. Nguyen, K.V. Kissinger, *et al.* 2002. Myocardial drug distribution pattern following intrapericardial delivery: an MRI analysis. *J. Cardiovasc. Magn. Reson.* **4:** 311–316.
18. Laham, R.J., M. Rezaee, M. Post, *et al.* 2003. Intrapericardial administration of basic fibroblast growth factor: myocardial and tissue distribution and comparison with intracoronary and intravenous administration. *Catheter Cardiovasc. Interv.* **58:** 375–381.
19. Kraehenbuehl, T.P., L.S. Ferreira, P. Zammaretti, *et al.* 2009. Cell-responsive hydrogel for encapsulation of vascular cells. *Biomaterials* **30:** 4318–4324.
20. Chiu, L.L.Y. & M. Radisic. 2011. Controlled-release of thymosin beta 4 using collagen-chitosan composite hydrogels promotes epicardial cell migration and angiogenesis. *J. Control. Release* **155:** 376–385.
21. Iwakura, A., M. Fujita, K. Kataoka, *et al.* 2003. Intramyocardial sustained delivery of basic fibroblast growth factor improves angiogenesis and ventricular function in a rat infarct model. *Heart Vessels* **18:** 93–99.
22. Liu, Y., L. Sun, Y. Huan, *et al.* 2006. Effects of basic fibroblast growth factor microspheres on angiogenesis in ischemic myocardium and cardiac function: analysis with dobutamine cardiovascular magnetic resonance tagging. *Eur. J. Cardiothorac. Surg.* **30:** 103–107.
23. Wei, H.J., H.H. Yang, C.H. Chen, *et al.* 2007. Gelatin microspheres encapsulated with a nonpeptide angiogenic agent, ginsenoside Rg1, for intramyocardial injection in a rat model with infarcted myocardium. *J. Control Release* **120:** 27–34.
24. Yamamoto, T., N. Suto, T. Okubo, *et al.* 2001. Intramyocardial delivery of basic fibroblast growth factor-impregnated gelatin hydrogel microspheres enhances collateral circulation to infarcted canine myocardium. *Jpn Circ. J.* **65:** 439–444.
25. Hsieh, P.C., M.E. Davis, J. Gannon, *et al.* 2006. Controlled delivery of PDGF-BB for myocardial protection using injectable self-assembling peptide nanofibers. *J. Clin. Invest.* **116:** 237–248.
26. Hao, X., E.A. Silva, A. Mansson-Broberg, *et al.* 2007. Angiogenic effects of sequential release of VEGF-A165 and PDGF-BB with alginate hydrogels after myocardial infarction. *Cardiovasc. Res.* **75:** 178–185.
27. Laham, R.J., F.W. Sellke, E.R. Edelman, *et al.* 1999. Local perivascular delivery of basic fibroblast growth factor in patients undergoing coronary bypass surgery: results of a phase I randomized, double-blind, placebo-controlled trial. *Circulation* **100:** 1865–1871.
28. Jain, R.A. 2000. The manufacturing techniques of various drug loaded biodegradable poly(lactide-co-glycolide) (PLGA) devices. *Biomaterials* **21:** 2475–2490.
29. Alonso, M.J., S. Cohen, T.G. Park, *et al.* 1993. Determinants of release rate of tetanus vaccine from polyester microspheres. *Pharm. Res.* **10**: 945–953.
30. Schwendeman, S.P., H.R. Costantino, R.K. Gupta, *et al.* 1996. Strategies for stabilising tetanus toxoid towards the development of a single-dose tetanus vaccine. *Dev. Biol. Stand.* **87:** 293–306.
31. Camarata, P.J., R. Suryanarayanan, D.A. Turner, *et al.* 1992. Sustained release of nerve growth factor from biodegradable polymer microspheres. *Neurosurgery* **30:** 313–319.
32. Cleland, J.L., E.T. Duenas, A. Park, *et al.* 2001. Development of poly-(D,L-lactide–coglycolide) microsphere formulations containing recombinant human vascular endothelial growth factor to promote local angiogenesis. *J. Control Release* **72:** 13–24.

33. Kostanski, J.W., B.A. Dani, B. Schrier & P.P. DeLuca. 2000. Effect of the concurrent LHRH antagonist administration with a LHRH superagonist in rats. *Pharm. Res.* **17:** 445–450.
34. Okada, H., Y. Doken, Y. Ogawa & H. Toguchi. 1994. Preparation of three-month depot injectable microspheres of leuprorelin acetate using biodegradable polymers. *Pharm. Res.* **11:** 1143–1147.
35. Athanasiou, K.A., G.G. Niederauer & C.M. Agrawal. 1996. Sterilization, toxicity, biocompatibility and clinical applications of polylactic acid/polyglycolic acid copolymers. *Biomaterials* **17:** 93–102.
36. Fournier, E., C. Passirani, C.N. Montero-Menei & J.P. Benoit. 2003. Biocompatibility of implantable synthetic polymeric drug carriers: focus on brain biocompatibility. *Biomaterials* **24:** 3311–3331.
37. Ratner, B. & A. Hoffman. 2004. *Biomaterials Science: An Introduction to Materials in Medicine*. 2nd ed. Elsevier Inc. San Diego, California.
38. Carrascosa, C., I. Torres-Aleman, C. Lopez-Lopez, *et al.* 2004. Microspheres containing insulin-like growth factor I for treatment of chronic neurodegeneration. *Biomaterials* **25:** 707–714.
39. Cho, K.Y., S.H. Choi, C.H. Kim, *et al.* 2001. Protein release microparticles based on the blend of poly(D,L-lactic-co-glycolic acid) and oligo-ethylene glycol grafted poly(L-lactide). *J. Control Release* **76:** 275–284.
40. Ehtezazi, T., C. Washington & C.D. Melia. 1999. Determination of the internal morphology of poly (D,L-lactide) microspheres using stereological methods. *J. Control Release* **57:** 301–314.
41. Rosa, G.D., R. Iommelli, M.I. La Rotonda, *et al.* 2000. Influence of the co-encapsulation of different non-ionic surfactants on the properties of PLGA insulin-loaded microspheres. *J. Control Release* **69:** 283–295.
42. Ikada, Y. & H. Tsuji. 2000. Biodegradable polyesters for medical and ecological applications. *Macromol Rapid Comm.* **21:** 117–132.
43. Kim, H.K. & T.G. Park. 2004. Comparative study on sustained release of human growth hormone from semi-crystalline poly(L-lactic acid) and amorphous poly(D,L-lactic-co-glycolic acid) microspheres: morphological effect on protein release. *J. Control Release* **98:** 115–125.
44. Cui, F., D. Cun, A. Tao, *et al.* 2005. Preparation and characterization of melittin-loaded poly (DL-lactic acid) or poly(DL-lactic-co-glycolic acid) microspheres made by the double emulsion method. *J. Control Release* **107:** 310–319.
45. Pean, J.M., M.C. Venier-Julienne, F. Boury, *et al.* 1998. NGF release from poly(D,L-lactide-co-glycolide) microspheres. Effect of some formulation parameters on encapsulated NGF stability. *J. Control Release* **56:** 175–187.
46. Wei, G., G.J. Pettway, L.K. McCauley & P.X. Ma. 2004. The release profiles and bioactivity of parathyroid hormone from poly(lactic-co-glycolic acid) microspheres. *Biomaterials* **25:** 345–352.
47. Diaz, R.V., M. Llabres & C. Evora. 1999. One-month sustained release microspheres of 125I-bovine calcitonin. In vitro-in vivo studies. *J. Control Release* **59:** 55–62.
48. Cleek, R., K. Ting, S.G. Eskin & A. Mikos. 1997. Microparticles of poly(dl-lactic-co-glycolic acid)/poly(ethylene glycol) blends for controlled drug delivery. *J. Control Release.* **48:** 259–268.
49. Ruan, G., S.S. Feng & Q.T. Li. 2002. Effects of material hydrophobicity on physical properties of polymeric microspheres formed by double emulsion process. *J. Control Release* **84:** 151–160.
50. Ungaro, F., M. Biondi, I. d'Angelo, *et al.* 2006. Microsphere-integrated collagen scaffolds for tissue engineering: effect of microsphere formulation and scaffold properties on protein release kinetics. *J. Control Release* **113:** 128–136.
51. Wei, G., L.F. Lu & W.Y. Lu. 2007. Stabilization of recombinant human growth hormone against emulsification-induced aggregation by Pluronic surfactants during microencapsulation. *Int. J. Pharm.* **338:** 125–132.
52. Hof, R.P., A. Hof, R. Salzmann & F. Wyler. 1981. Trapping and intramyocardial distribution of microspheres with different diameters in cat and rabbit hearts in vitro. *Basic Res. Cardiol.* **76:** 630–638.
53. Grizzi, I., H. Garreau, S. Li & M. Vert. 1995. Hydrolytic degradation of devices based on poly(DL-lactic acid) size-dependence. *Biomaterials* **16:** 305–311.
54. Vert, M., J. Mauduit & S. Li. 1994. Biodegradation of PLA/GA polymers: increasing complexity. *Biomaterials* **15:** 1209–1213.
55. Shenderova, A., A. Ding & S. Schwendeman. 2004. Potentiometric method for determination of microclimate pH in Poly(lactic-co-glycolic acid) films. *Macromolecules* **37:** 10052–10058.
56. Lucke, A., J. Kiermaier & A. Gopferich. 2002. Peptide acylation by poly(alpha-hydroxy esters). *Pharm. Res.* **19:** 175–181.
57. Sophocleous, A.M., Y. Zhang & S.P. Schwendeman. 2009. A new class of inhibitors of peptide sorption and acylation in PLGA. *J. Control Release* **137:** 179–184.
58. Watson, R.M., D.R. Markle, *et al.* 1984. Transmural pH gradient in canine myocardial ischemia. *Am. J. Physiol.* **246:** H232–H238.
59. Sun, Y. 2007. Oxidative stress and cardiac repair/remodeling following infarction. *Am. J. Med. Sci.* **334:** 197–205.

Ann. N.Y. Acad. Sci. ISSN 0077-8923

ANNALS OF THE NEW YORK ACADEMY OF SCIENCES
Issue: *Thymosins in Health and Disease*

Corrigendum for Ann. N.Y. Acad. Sci. 2005. 1051: 779–786

Levy, Y., Y. Uziel, G. Zandman, P. Rotman, H. Amital, Y. Sherer, P. Langevitz, B. Goldman & Y. Shoenfeld. 2005. Response of vasculitic peripheral neuropathy to intravenous immunoglobulin. *Ann. N.Y. Acad. Sci.* **1051:** 779–786.

The above-cited *Annals* paper was a part of the conference proceedings of the Fourth International Congress on Autoimmunity (2004) that reported follow-up data related to a previously published paper in *Annals of the Rheumatic Diseases* (Intravenous immunoglobulins in peripheral neuropathy associated with vasculitis. 2003. 62: 1221–1223). Although citing the *Annals of the Rheumatic Diseases* paper, the *Annals* paper should have more clearly described the relationship between the two papers. We apologize to the readers and editors of *Annals of the Rheumatic Diseases.*

doi: 10.1111/j.1749-6632.2012.06792.x

ANNALS OF THE NEW YORK ACADEMY OF SCIENCES
Issue: *Thymosins in Health and Disease*

Corrigendum for Ann. N.Y. Acad. Sci. 2012. 1254: 57–65

Soler-Botija, C., J.R. Bagó & A. Bayes-Genis. 2012. A bird's-eye view of cell therapy and tissue engineering for cardiac regeneration. *Ann. N.Y. Acad. Sci.* **1254:** 57–65.

The following acknowledgment was inadvertently omitted from the above-mentioned article.

Researchers working on the project that led to these results received funding from the European Union Seventh Framework Programme (FP7/2007–2013) under Grant Agreement No. 229239. The members of the RECATABI consortium are Carlos Semino, Juan Carlos Chachques, Manuel Monleón-Pradas, Antoni Bayes-Genis, Nicole zur Nieden, and Creaspine SAS.

doi: 10.1111/j.1749-6632.2012.06793.x